I0043510

John Hopkinson

Original Papers on Dynamo Machinery and Allied Subjects

John Hopkinson

Original Papers on Dynamo Machinery and Allied Subjects

ISBN/EAN: 9783337007560

Printed in Europe, USA, Canada, Australia, Japan

Cover: Foto ©berggeist007 / pixelio.de

More available books at **www.hansebooks.com**

ORIGINAL PAPERS

ON

DYNAMO MACHINERY

AND ALLIED SUBJECTS.

BY

JOHN HOPKINSON, M.A., D.Sc., F.R.S.

NEW YORK:

THE W. J. JOHNSTON COMPANY, LIMITED,

41 PARK ROW (TIMES BUILDING).

LONDON:

WHITTAKER & CO.,

2, WHITEHART STREET, PATERNOSTER SQUARE.

1893.

AUTHORIZED AMERICAN EDITION.

PREFACE.

THE following short collection of papers includes all that I have written of an original character on electrotechnical subjects. Here and there errors have been corrected; otherwise the papers are republished exactly as they first appeared. The chronological order is not strictly adhered to. The papers are arranged rather according to subject. Thus, five papers relating wholly or in part to the continuous current dynamo come first; then follow four on converters; lastly, there are a note on the theory of alternate current machines and a paper on the applications of electricity to lighthouses.

The motive of this publication has been that I have understood that one or two of these papers are out of print and are not so accessible to American readers as an author who very greatly values the good opinion of American electrical engineers would desire.

J. HOPKINSON.

LONDON, September, 1892.

CONTENTS.

ORIGINAL PAPERS

ON

DYNAMO MACHINERY

AND ALLIED SUBJECTS.

ON ELECTRIC LIGHTING.

FIRST PAPER.

DURING the last year much has been written and much information communicated concerning the production of light from mechanical power by means of an electric current. The major portion of what has appeared has been either descriptive of particular machines for producing the current, and of lamps for manifesting a portion of its energy as light, or a statement of practical results connecting the light obtained with the power applied and the money expended in producing it.

While fully appreciating the present value of such information, the author has felt that it did not tell all that was interesting or practically useful to know. It is desirable to know what the various machines can do with varied and known resistances in the circuit, and with varied speeds of rotation; and what amount of power is absorbed in each case. It is a question of interest whether

7

a machine intended for one light can or cannot produce two in the same circuit, and if not, why not; whether a machine such as the Wallace-Farmer, intended as it is for many lights, will give economical results when used for one; and so on. It is clear that the attempt to examine all separate combinations of so many variables would be hopeless, and that the work must be systematized.

The mechanical energy communicated by the steam engine or other motor is not immediately converted into the energy of heat, but is first converted into the energy of an electric current in a conducting circuit; of this a portion only becomes localized as heat between the carbons of the electric arc; and of this again a part only becomes sensible to the eye as light. The whole of what we need to know may be more easily ascertained and more shortly expressed if the inquiry is divided into two parts: (a) What current will a machine produce under various conditions of circuit, and at what expenditure of mechanical power? (b) Having given the electric conditions under which the arc is placed, no matter how these conditions are produced, what light will be obtained therefrom? Parts of the subject have been treated more or less in this sense by Edlund (*Pogg. Ann.*, 1867 and 1868), Houston and Thomson in America, Mascart (*Journal de Physique,* March, 1878), Abney (*Proceedings of the Royal Society,* 1878), Trowbridge (*Philosophical Magazine,* March, 1879), Schwendler (Report on Electric Light Experiments), etc., but not so completely that nothing remains to be done; nor does the author doubt that a great deal of information is in the hands of makers of machines, which they have not thought it necessary to make known. The present communication is limited to an account of some experiments on the production of currents by a Siemens medium-sized machine; that is, the machine which is advertised to produce a light of 6,000 candles by an expenditure of 3½ horse power.

All the machines for converting mechanical power into an electric current consist ultimately of a conducting wire moving in a magnetic field; and approximately the electromotive force of the machine will be proportional to the velocity with which the circuit moves through the field, and to the intensity of the field. In general the intensity of the field is not constant; and in such machines as the Siemens and the ordinary Gramme machine it may be regarded as a function of the current passing. We must learn what this function is for the machine in question; or—which comes to exactly the same thing, and is better so long as the facts are merely the result of experiment—we must construct a curve in which the abscissæ represent the intensities of currents passing, and the ordinates the corresponding electromotive forces for a given speed of rotation. But the power of a current, that is, its energy per second, is the product of the electromotive force and the intensity, or, in the case of the curve, the product of the ordinate and the abscissa; this is in all cases less than the power required to drive the machine, and the ratio between the two may fairly be called the efficiency of the machine.

The object of the inquiry may perhaps be made clearer by an illustration. Consider the case of a pump forcing water through a pipe against friction; then the electric current corresponds to the volume of water passing per

second, and the electromotive force to the difference of
pressure on the two sides of the pump; and just as the
product of pressure and volume per second is power, so
the product of electromotive force and current is power,
which is directly comparable with the power expended in
driving the machine or the pump, as the case may be.
The peculiarity of the so-called dynamo-electric machine
lies in this, that what corresponds to the difference of
pressure (the electromotive force) depends directly on
what corresponds to the volume passing (the current).

Each experiment requires the determination of the
speed, the driving power, the resistances in circuit, and
the current passing; or of the difference of potential between the two ends of a known resistance of the circuit.

The apparatus employed by the author was arranged,
not alone with an aim to accuracy, but in part to make use
of such instruments as he happened to possess or could
easily construct, and in part with a view to ready erection
and transportation. Much more accurate results may be
obtained by any one who will arrange apparatus with a single
aim to attain the greatest accuracy possible. The author's
apparatus will, however, be briefly described, that others
may form their own opinion of the importance of the various sources of error.

The speed counter was that supplied with the electric
machine.

Concerning the steam engine nothing need be said, save
that its speed was maintained very constant by means of a
governor, shown in Fig. 1, specially arranged for great
sensibility. By placing the joint A above the joint B, instead of below it, as in Porter's governor, any degree of

sensibility up to instability may be obtained. The speed
was varied by means of a weight and a spring attached to
a lever on the throttle valve
spindle. The ungainly ap-
pearance of this governor could
easily be remedied by any one
proposing to manufacture it.

The power is transmitted
from the engine to a countershaft by means of a strap, and
by a second strap from the
countershaft to the pulley of
the electric machine. On this
second strap is the dynamometer shown in Fig. 2.

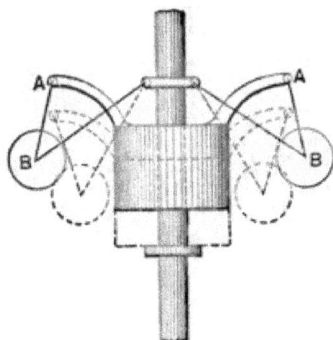

FIG. 1.—GOVERNOR.

This dynamometer has for some time been used by
Messrs. Siemens, and was also used by Mr. Schwendler; its
invention is due to Herr von Hefner-Alteneck. A is the
driving pulley; B the pulley of the electric machine; $C C$
are a pair of loose pulleys between which the strap passes;
these are carried in a double triangular frame, which can
turn about a bar D. This bar might form part of a permanent structure; but in order to place the dynamometer
readily on any strap, the bar was in this case provided with
eyes at either end, and secured in position by six or eight
ropes. This plan answers well, as there is very little stress
on the bar. Immediately above the pulleys $C C$ a cord
leads from the frame through a Salter spring balance over
snatch blocks to a back balance weight; the tension of this
cord is read on the spring balance. At first the spring
balance was omitted, and the weight at the end of the cord

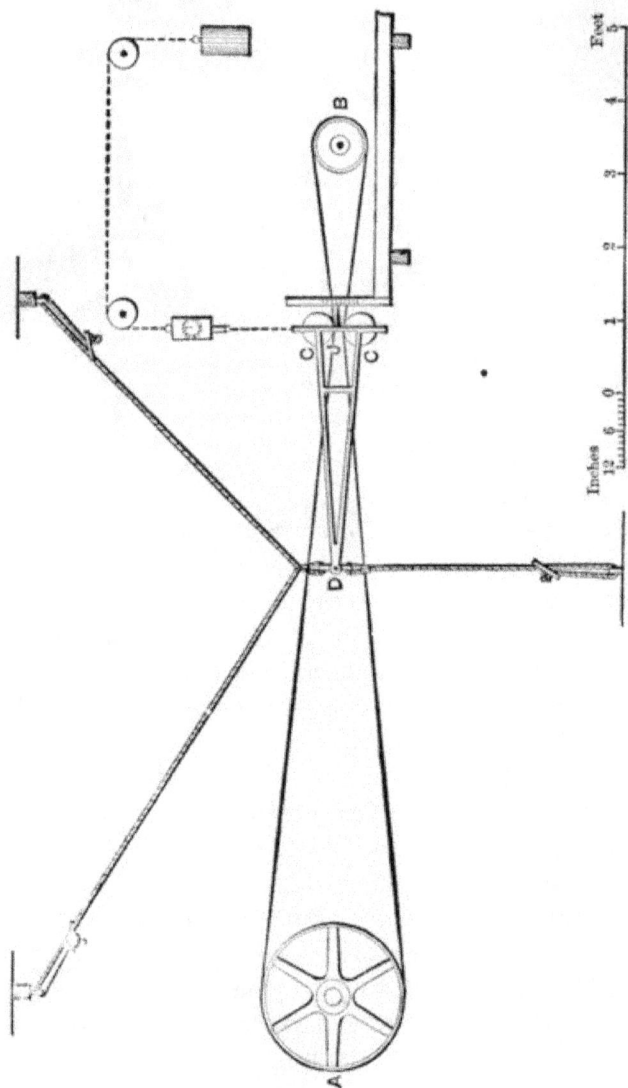

FIG. 2.—DYNAMOMETER FOR MEASURING POWER TRANSMITTED THROUGH BELT.

was observed; but the friction of the snatch block pulleys was found objectionable. The pulley frame carries a pointer, which is adjusted so as to coincide with a datum mark when the line $A\,B$ bisects the distance between the loose pulleys. Let W' be the tension of the cord required to bring the pulley frame to its standard position when no work is being transmitted, W'' the tension which is required to bring the pointer back to the datum mark when an observation is made, and let $W = W' - W''$. Let T', T'' be the tensions on the tight and slack halves of the strap; R_1, R_2, r the radii of the pulleys A, B and C, plus half the thickness of the strap; c_1, c_2 the distances $A\,J$, $J\,B$; $2d$ the distance apart of the centres C, C; a_1, a_2 the inclinations of the two parts of the strap, on either side of C, C, to the line $A\,B$. Then

$$(T' - T'')(\sin a_1 + \sin a_2) = W;$$

and

$$\sin a_1 = \frac{R_1 + r - d}{c_1} + \frac{d}{2c_1}\left(\frac{R_1 + r - d}{c_1}\right)^2,$$

$$\sin a_2 = \frac{R_2 + r - d}{c_2} + \frac{d}{2c_2}\left(\frac{R_2 + r - d}{c_2}\right)^2,$$

very nearly.

The value of $T' - T''$ and the velocity of rotation of the electric machine being known, the power received by it is readily obtained, expressed in gram-centimetres per second. Multiplying by 981, the value of gravity in centimetres and seconds, the power is then expressed in ergs* per

---

* The dyne is the force which will in one second impart to one gram a velocity of one centimetre per second, and an erg is the work done by a dyne working through a centimetre; a horse power may be taken as three-quarters

second, and is ready for comparison with the **results of** the electrical experiments.

As already stated, the dynamo-electric **machine in** the present case was a Siemens medium size; the armature coil has fifty-six divisions, and the brushes are single, **not** divided, **that is, each** brush is in connection with **one segment of the** commutator at any instant.

The leading wire is 100 yards of Siemens No. 90, con-**sisting of seven copper wires,** insulated with tape and india **rubber, and having a diameter of** about 9½ mm.

The method of determining the current is shown in the diagram, **Fig. 3.** The current is conveyed from the machine *A* through a set of coils of brass wire *c*, and in some cases through a resistance coil placed in a calorimeter *B*, and so back to the machine, the connections being made through cups of mercury excavated in a piece of wood *D*. The current passing may be ascertained by the heating of the calorimeter, or by measuring the difference of potential at the extremities of the resistance *c*, all the resistances of the circuit being supposed known. This difference of potential could of course be very easily measured by means of a quadrant electrometer; but, as the instrument had to be frequently removed, a galva-nometer appeared more convenient. The two points to be measured are connected to the ends of two series of resist-ance coils *a*, *b*. The galvanometer *G* is placed in a second derived current, passing from a junction in *a b* through a battery *H*, then through a set of high resistances *J* for

---

of an erg-ten per second, an erg-ten being 10⁷ ergs. See *Report of the Brit. Asso.*, 1873: and Everett "On the C. G. S. System of Units," published by the Physical Society.

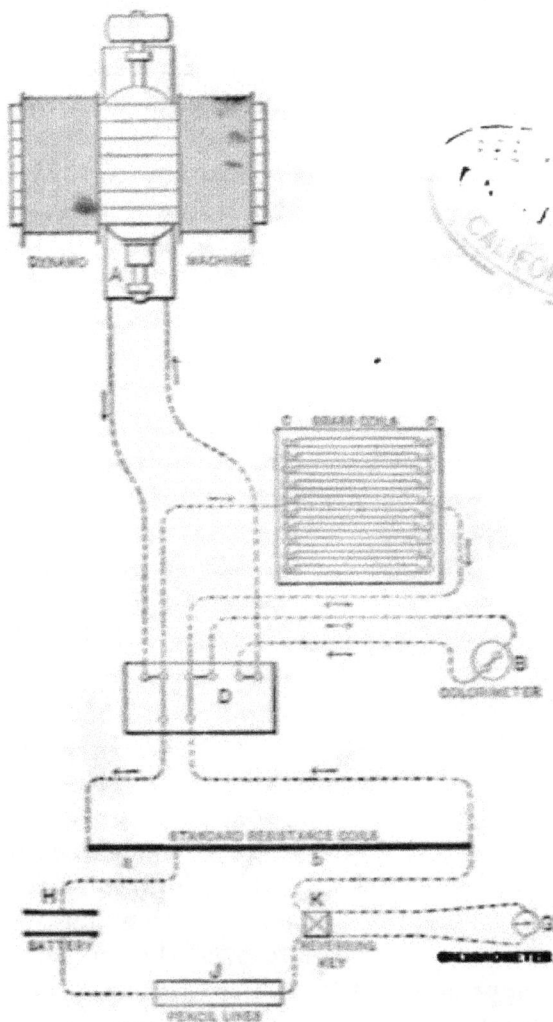

FIG. 2.—DIAGRAM OF APPARATUS.

adjusting sensibility, a reversing key K, the galvanometer G, the reversing key K again, and so to the other extremity of b. The electromotive force is ascertained by adjusting the resistance b so that the deflection of the galvanometer is nil.

The resistance coils c comprise ten coils of common brass wire, each wound round a couple of wooden uprights driven into a baseboard common to the set; each wire is about 60 metres long, and of No. 17 Birmingham wire gauge (.06 inch or $1\frac{1}{2}$ mm. diameter), weighing about 14.6 grams per metre. Each terminal is connected to a cup of mercury excavated in the baseboard, so that the coils can be placed in series or in parallel circuit at pleasure. The resistance of each coil being about 3 ohms, this set may be arranged to give resistances varying from 0.3 to 30 ohms.

The calorimeter B is a Siemens pyrometer with the top scale removed; a resistance coil of uncovered German silver wire nearly 2 m. long, $1\frac{1}{2}$ mm. in diameter, and having a resistance of about 0.2 ohm, is suspended within it from an ebonite cover, which also carries a little brass stirrer, and the calorimeter is filled with water to a level determined by the mark of a scriber. It was of course necessary to know the capacity of the calorimeter for heat. It was filled with warm water up to the mark, and the coil placed in position; 120 grams of water were then withdrawn, and the temperature of the calorimeter was observed to be 58.8° C.; after the lapse of one minute it was 58.3° C.; after a second minute 57.9° C.; 120 grams of cold water, temperature 13.3° C., were then suddenly introduced through a hole in the ebonite cover, and it was found that, two minutes after the reading of 57.9° C., the temperature was 50.0° C.; hence we may infer that the capacity of the calorimeter is equal to that of 740 grams of water. Two similar experiments at lower temperatures gave respectively the numbers 749 and 750. Estimating the capacity from the weight of the copper cylinder supplied with the pyrometer, it should be 747, to which must be added the capacity of the German silver wire and stirrer. Taking everything into consideration, 750 grams may be assumed as the most probable result.

The resistance coils a, b are of German silver, made by Messrs. Elliott Brothers; they are on the binary scale from $\frac{1}{8}$ ohm to 1,024 ohms. Separate coils were used, instead of a regular resistance box, because they were more readily applicable to any other purpose for which they might be required; and the binary scale was adopted, because the coils could at once be used as conductivity coils in parallel circuit, also on the binary scale. Each coil as supplied terminated in two stout copper legs; these were fitted with cups of india rubber tubing for mercury, whereby any connections whatever could readily be made. This arrangement, though rude, was very convenient, and perhaps even safer from error than a box with brass plugs to make the connections. By a slight alteration of the connections the whole was instantly available as a Wheatstone bridge to determine resistances.

The battery H is a single element of Daniell's battery, in which the sulphate of zinc solution floats on the sulphate of copper; its electromotive force is assumed to be $\frac{2}{3}$ volt.

The resistances J added in the battery circuit are pencil lines on glass, such as are described in the *Philosophical*

Magazine of February, 1879. Three were used, giving a range of sensibility approximately in the proportions 1, 25, 170, 700—the last figure being when all were short circuited; they are very useful in adjusting the resistance b so as to give no deflection of the galvanometer.

The reversing key K belongs to Sir W. Thomson's electrometer, and is quite suitable when high resistances and nil methods are used.

The galvanometer G is far more sensitive than necessary, and has a resistance of 7,000 ohms.

Preliminary to experiments on the current, determinations of resistances were made. The resistance of each brass coil c was first determined, to afford the means of calculating the value of this resistance in any subsequent experiment. When the ten coils were coupled in parallel circuit, the calculated resistance was 0.29 ohm, while 0.292 was obtained by direct measurement. The leading wire was then examined; the further ends being disconnected, the insulation resistance was found to be over 60,000 ohms; how much over, it was immaterial to learn. When the ends of the wire were connected, the resistance was found to be 0.129 ohm. The resistances in the dynamo-electric machine A were found to be as follows when cold: magnet coils, 0.156 and 0.152 respectively; armature coil, 0.324; total, 0.632 ohm. Direct examination was made of the whole machine in eight positions of the commutator, giving 0.643 ohm, with a maximum variation of 0.6 per cent. from the mean. After running the machine for some time the resistance was found to be 0.683, an increase which would be accounted for by a rise of temperature of 12° C. or thereabouts. The resistance of the calorimeter

B is 0.20 ohm, without its leading wire, which may be taken as 0.01. We have then in circuit three resistances which must be considered: (1) The resistance of the machine A and leading wire, assumed throughout as together 0.81 ohm, and denoted by c_1; (2) the resistance of the brass coils c, calculated from the several determinations, with the addition of 0.02 ohm, the resistance of the leading wire, and denoted by c_2; (3) when present, the resistance of the calorimeter B and leading wire, denoted by c_3.

Two approximate corrections were employed, and should be detailed. The first is the correction for the considerable heating of the resistance coils c. These were arranged in two sets of five each, five being in parallel circuit, and the two sets in series. The current from the machine, being about 7.4 webers in each wire, was passed for three or four minutes; the circuit was then broken, and the resistance c_2 was determined within one second of breaking circuit, when it was found to be about 5 per cent. greater than when cold. As the resistance was falling, the following was adopted as a rule of correction: square the current in a single wire, and increase the resistance c_2 by $\frac{1}{10}$ per cent. for every unit in the square. The second correction is due to the fact that the calorimeter was losing heat all the time it was being used. It was assumed that it loses 0.01° C. per minute for every 1° C. by which the temperature of the calorimeter exceeds that of the air; this correction is of course based on the experiment already mentioned.

The method of calculation may now be explained:—

R is the total resistance of the circuit in ohms, equal to $c_1 + c_2 + c_3$;

Q is the current passing in webers;

E is the electromotive force round the circuit in volts;

W_1 is the work per second converted into heat in the circuit, as determined by the galvanometer, measured in erg-tens per second;

W_2 is the work per second as determined by the calorimeter;

W_3 is the work per second as determined by the dynamometer, less the power required to drive the machine when the circuit is open;

$H P$ is the equivalent of W_3 in horse power;

n is the number of revolutions per minute of the armature.

As already mentioned, the standard resistance coils a, b are adjusted in each experiment so that the galvanometer gives no deflection, and the value of b is then noted. The values of c_1, c_2, c_3 are known from previous observations. Then

$$Q = \frac{9}{8} \times \frac{a+b}{b} \times \frac{1}{c_2},$$

$$E = Q \times R,$$

$$W_1 = E \times Q,$$

$$W_2 = \frac{R}{0.2} \times \left\{ \begin{array}{l} \text{mechanical equivalent of heat} \\ \text{generated per second in calorimeter.} \end{array} \right\}$$

The results of the experiments are given in the accompanying table.

TABLE I.—EXPERIMENTS ON SIEMENS DYNAMO-ELECTRIC MACHINE.

No. of Experiment.	Total Resistance of Circuit. R	Electric Current. Q	Electromotive Force. E	Work measured per second.				Revs. of Armature per minute. n	Position of Commutator Brush.
				By Galvanometer. W_1	By Calorimeter. W_2	By Dynamometer.			
	Ohms.	Webers.	Volts.	Erg-tens.	Erg-tens.	W_3 Erg-tens.	Horse Power. H.P.	Revs.	
1	1.025	0.0027	2.72	....	....	....	....	720	
2	8.3	0.48	3.95	0.0019	....	....	....	"	
3	5.33	1.45	7.73	0.0012	....	0.042	0.056	"	
4	4.07	16.8	68.4	1.149	1.140	1.179	1.59	"	
5	3.88	18.2	70.6	1.285	....	1.263	1.68	"	
6	3.205	24.8	79.5	1.972	2.158	2.106	2.81	"	Brush in original position, as supplied from maker.
7	3.025	26.8	81.1	2.174	....	2.392	3.19	"	
8	2.62	32.2	84.4	2.718	2.888	2.790	3.71	"	
9	2.43	34.5	83.8	2.894	....	3.370	4.49	"	
10	2.28	37.1	84.6	3.138	2.903	3.538	4.72	"	
11	2.08	42.0	87.4	3.671	....	3.960	5.28	"	(Strap slipping.)
12	1.345	64.0	86.1	5.510	5.349	5.777	7.70	654	
13	2.08	41.1	85.5	3.514	....	3.790	5.05	698	13° extra lead of brush.
14	2.07	36.0	74.5	2.692	....	2.952	3.80	696	
15	2.09	42.7	89.2	3.809	....	4.233	5.64	708	8°
16	2.09	41.7	87.2	3.636	....	4.135	5.51	713	5° less lead of brush than in original position.
17	2.10	45.0	94.5	4.252	....	4.810	6.41	759	5°
18	2.08	40.8	84.9	3.464	....	4.010	5.35	696	5°
19	2.06	35.1	72.3	2.538	....	2.672	3.56	586	5°

A power of 0.21 erg-ten, or 0.28 horse power, was required to drive the machine at 720 revolutions on open circuit. An examination of the table shows that the efficiency of the machine is about 90 per cent., exclusive of friction. Comparing experiments 11 and 13, and also the last four experiments, it is seen that the electromotive force is proportional to the speed of rotation within the errors of observation. Experiments 14, 15, and 16 were intended to ascertain the effect of displacing the commutator brushes.

The principal object of the experiments was to ascertain how the electromotive force depended on the current. This relation is represented by the curve shown in Fig. 4,

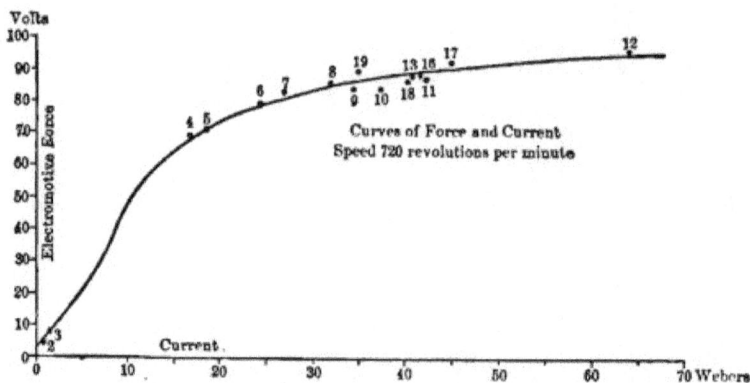

FIG. 4.—CURVE OF FORCE AND CURRENT.

in which the abscissæ represent the currents flowing, or the values of Q in the table, and the ordinates the electromotive forces, or the values of E reduced to a speed of 720 revolutions per minute. The curve may also be taken to represent the intensity of the magnetic field. It will be remarked that there is a point of inflection in the curve somewhere near the origin. The experiments 1 to 5 indicate that this is the true form of the curve, and it is confirmed in a remarkable manner by a special experiment. A resistance intermediate between 5½ and 4 (experiments 3 and 4) was used in circuit, and E and Q were determined in two different ways: first, by starting with an open circuit, which was then closed; secondly, by starting with a portion of the resistance short circuited, and a very powerful current passing, and then breaking the short circuit. It was found that E and Q were four times as great in the latter case as in the former. Unfortunately the numbers are not sufficiently accurate to be given, as the solutions of the standard battery had become mixed.

The curve really gives a great deal more information than appears at first sight. It will determine what current will flow at any given speed of rotation of the machine, and under any conditions of the circuit, whether of resistances or of opposed electromotive forces. It will also give very approximate indications of the corresponding curve for other machines of the same configuration, but in which the number of times the wire passes round the electromagnet or the armature is different.

It will be well to compare these results with those obtained by others. M. Mascart worked on a Gramme machine with comparatively low currents: he represents his results approximately by the formula

$$E = n\,(a + b\,Q),$$

where a and b are constants. This corresponds to the rapidly rising part of the curve in Fig. 4. Mr. Trowbridge

with a Siemens machine obtained a maximum efficiency of
76 per cent., and states that the machine was running
below its normal velocity. Mr. Schwendler's results, when
fully published, will probably be found to be the most
complete and most accurate existing. In the *précis* he
states that the loss of power with a Siemens machine in
producing currents of over 20 webers is 12 per cent.
Now, taking the author's experiments 4 to 19, the mean
value of W_1 is 3.027 erg-tens and of W_2 3.304; adding to
the latter 0.21, the power required to drive the machine
when no current passes, it appears that 13.8 per cent. of
the power applied is wasted. Again, taking experiments
4, 6, 8, 10, and 12, the mean value of W_2 is 2.888 erg-tens
and of W_1 3.076, indicating a waste of power amounting to
12 per cent. Of this, as already stated, 0.21 erg-ten, or
0.28 horse power, is accounted for by friction of the
journals and commutator brush; the remainder is expended in local currents, or by loss of kinetic energy of
current when sparks occur at the commutator.

According to Weber's theory of induced magnetism, as
set forth in Maxwell's "Electricity and Magnetism,"
vol. II., if X be the magnetizing force and I the intensity
of magnetization,

$$I = \frac{2}{3} a \frac{X}{b}, \text{ until } X \text{ rises to the value } b,$$

and
$$I = a \left(1 - \frac{1}{3} \frac{b^2}{X^2}\right), \text{ if } X > b,$$

where a and b are constants. We should naturally expect
that a similar formula would be approximately applicable
to dynamo-electric machines.

In the present experiments, let I be the electromotive force, X the current passing, and assume a to be 60 and b to be 15; we then obtain results not far from those of experiment. The capacity of any continuous current machine may thus be shortly stated by giving the values of a and b; or, which comes to the same thing, by stating the electromotive force at a given speed when the current is as great as possible, and also the total resistance through which the machine will exert an electromotive force two-thirds of this greatest electromotive force. To this should be added a statement of the resistance of the machine, and of the power it absorbs, with known conditions of the circuit.

The author has not yet tried any quantitative experiments with the electric light, but hopes shortly to do so. In the meantime he would remark that, as the lamp is usually adjusted, only half the energy of the current appears in the arc, or 44 per cent. of the energy transmitted to the machine by the strap.

In conclusion the author would express the obligation he is under to Messrs. Chance Brothers & Co., on account of the facilities he has enjoyed for making these experiments at their works. It may be mentioned that one principal object of the research of which this is a beginning is to obtain a minute knowledge of the electric light, with a view to lighthouse illumination.

ON ELECTRIC LIGHTING.

SECOND PAPER.

Dynamo-Electric Machines.—Since the date of the author's former paper in April, 1879, other observers have published the results of experiments similar to those described by him. It may be well to exhibit some of these

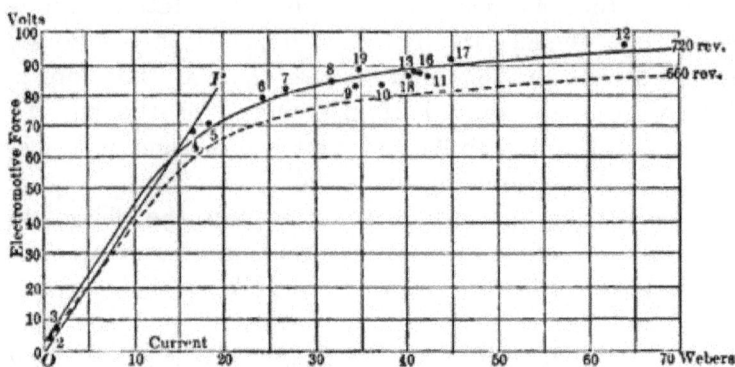

FIG. 5.—CURVES OF ELECTROMOTIVE FORCE AND CURRENT OF SIEMENS MEDIUM SIZE MACHINE.

results reduced to the form he has adopted, namely, a curve, such as that previously shown in Fig. 4 of the preceding paper, and now reproduced, with slight alterations, in Fig. 5. Here any abscissa represents a current passing through the dynamo-electric machine, and the corresponding ordinate represents the electromotive force of the machine for a certain speed of revolution, when that current is passing through it. It will be found (1) that with varying speed the ordinate, or electromotive force, corresponding to any abscissa or current is proportional to the speed; (2) that the electromotive force does not increase indefinitely with increasing current, but that the curve approaches an asymptote; (3) that the earlier part of the curve is, roughly speaking, a straight line, until the current attains a certain value, and that at that point the electromotive force has reached about two-thirds of its maximum value. When the current is such that the electromotive force is not more than two-thirds of its maximum, a very small change in the resistance with speed of engine constant, or in the speed of the engine with resistance constant, causes a great change in the current. For this reason the greatest of these currents, which is that corresponding to the point where the curve breaks away from a straight line, and which is the same for all speeds of revolution, since the curves for different speeds differ only in the scale of ordinates, may be called the " critical current " of the machine. The effect of a change of speed is exhibited in Fig. 5, where the lower dotted line represents the curve for a speed of 660 revolutions per minute, instead of 720. The resistance, varying as $\dfrac{\text{electromotive force}}{\text{current}}$, is given by the slope of the line $O\,P$. But since the resistance is constant, the slope of this line must be constant; and it will be seen that it cuts the upper curve at a point corresponding to a current of 15 webers, and the lower at a point corresponding to a current of 5 webers only.

In Germany, Auerbach and Meyer (*Wiedemann Annalen,*
Nov., 1879) have experimented fully on a Gramme machine
at various speeds, and with various external resistances.
The resistance of the machine was 0.97 ohm. Their results
are summarized in a table at the end of their paper, which
gives the current passing, with resistances in circuit from
1.75 to 200 Siemens units, and at speeds from 20 to 800
revolutions per minute. In the accompanying diagram,
Fig. 6, the curve *G* expresses the relation between electro-

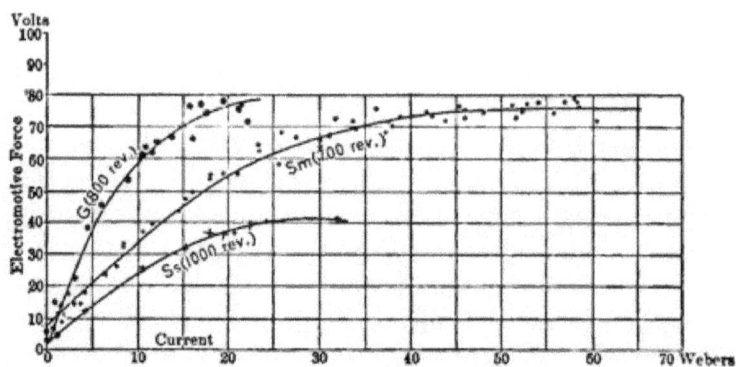

FIG. 6.—CURVES OF ELECTROMOTIVE FORCE AND CURRENT: GRAMME MACHINE,
G; SIEMENS MEDIUM, SM; SIEMENS SMALLEST, SS.

motive force and current, as deduced from some of their
observations; the points marked are plotted from their
table, making allowance, where necessary, for difference in
speed. The curve, as actually constructed, is for a speed
of 800 revolutions: at this speed it will be seen that the
maximum electromotive force is about 76 volts; and the
critical current, corresponding to a force of about 51 volts,
is 6.5 webers, with a total resistance of 7.8 ohms. Up to
this point there will be great instability, exactly as was the

case in the Siemens machine examined by the author, where the resistance was 4 ohms and the speed 720 revolutions.

The results of an elaborate series of experiments on certain dynamo-electric machines have recently been presented to the Royal Society by Dr. Siemens. One of the machines examined was an ordinary medium sized machine, substantially similar to that tried by the author in 1879. It is described as having 24 divisions of the commutator; 336 coils on the armature, with a resistance of 0.4014 Siemens unit; and 512 coils on the magnets, with a resistance of 0.3065: making a total resistance of 0.7079 Siemens unit $= 0.6654$ ohm. The curve $S\,m$, Fig. 6, gives the relation of electromotive force and current, reduced to a speed of 700 revolutions per minute, the actual speeds ranging from 450 to 800 revolutions. The maximum electromotive force appears to be probably 76 volts, and the critical current 15 webers; which is the same as in the author's first experiments on a similar machine.

In the summer of 1879 the author examined a Siemens machine of the smallest size. This machine is generally sold as an exciter for their alternate current machine. It has an internal resistance of 0.74 ohm, of which 0.395 is in the armature or helix. The machine is marked to run at 1,130 revolutions per minute. The following Table II. gives, for a speed of 1,000 revolutions, the total resistance, current, electromotive force and horse power developed as current. The horse power expended was not determined.

The curve $S\,s$, Fig. 6, gives as usual the relations of electromotive force and current. From this curve it will be seen that the critical current is 11.2 webers, and the maximum electromotive force, at the speed of 1,000 revolutions, is

TABLE II.—EXPERIMENTS ON SMALLEST-SIZED SIEMENS DYNAMO-
ELECTRIC MACHINE.

Resistance.	Electric Current.	Electromotive Force.	Horse Power Developed as Current.
2.631 ohms.	5.10 webers.	13.2 volts.	0.09 H. P.
2.221 "	12.15 "	27.0 "	0.44 "
1.967 "	17.0 "	33.6 "	0.76 "
1.784 "	20.4 "	36.4 "	0.99 "
1.668 "	22.3 "	37.2 "	1.11 "
1.579 "	23.2 "	36.6 "	1.14 "
1.503 "	25.6 "	39.3 "	1.34 "
1.440 "	27.8 "	40.0 "	1.49 "
1.145 "	36.2 "	41.5 "	2.00 "

about 42 volts.   The determinations for this machine were
made in exactly the same manner as in the experiments
on the medium sized machine, using the galvanometer, but
omitting the experiment with the calorimeter (compare
Table I., p. 21).

The time required to develop the current in a Gramme
machine has been examined by Herwig (Wiedemann, June,
1879).   He established the following facts for the machine
he examined : A reversed current, having an electromotive
force of 0.9 Grove cell, sufficed to destroy the residual
magnetism of the electromagnets.   If the residual magnetism was as far as possible reduced, it took a much longer
time to get up the current than when the machine was in
its usual state.   A longer time was required to get up the
current when the external resistance was great than when
it was small.   With ordinary resistance the current required
from ¾ second to 1 second to attain its maximum.

Brightness of the Electric Arc.—The measurement of
the light emitted by an electric arc presents certain peculiar
difficulties.   The light itself is of a different color from

that of a standard candle, in terms of which it is usual to express luminous intensities. The statement, without qualification, that a certain electric lamp and machine give a light of a specified number of candles, is therefore wanting in definite meaning. A red light cannot with propriety be said to be any particular multiple of a green light; nor can one light which is a mixture of colors be said with strictness to be a multiple of another, unless the proportions of the colors in the two cases are the same. Captain Abney (*Proceedings of the Royal Society*, March 7, 1878, p. 157) has given the results of measurements of the red, blue, and actinic light of electric arcs in terms of the red, blue, and actinic light of a standard candle. The fact that the electric light is a very different mixture of rays from the light of gas or of a candle has long been known, but has been ignored in statements intended for practical purposes.

Again, the emission of rays from the heated carbons and arc is by no means the same in all directions. Determinations have been made in Paris of the intensity in different directions, in particular cases. If the measurement is made in a horizontal direction, a very small obliquity in the crater of the positive carbon will throw the light much more on one side than on the other, causing great discordance in the results obtained.

If the electric light be compared directly with a standard candle, a dark chamber of great length is needed—a convenience not always attainable. In the experiments made at the South Foreland by Dr. Tyndall and Mr. Douglass, an intermediate standard was employed; the electric light was measured in terms of a large oil lamp, and this latter was frequently compared with a standard candle.

Other engagements have prevented the author from fairly attacking these difficulties; but since May, 1879, he has had in occasional use a photometer with which powerful lights can be measured in .moderate space.   This photometer is shown in Figs. 7 and 8, and an enlargement

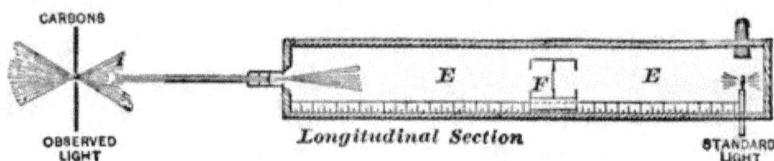

FIG. 7.—PHOTOMETER FOR POWERFUL LIGHTS.

of the field piece in Fig. 9.   A convex lens A, of short focus, forms an image at B of the powerful source of light which it is desired to examine.   The intensity of the light from this image will be less than that of the actual source by a calculable amount; and when the distance of the lens from the light is suitable, the reduction

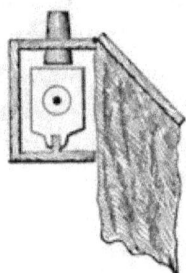

FIG. 8.—TRANSVERSE SECTION OF PHOTOMETER.

is such that the reduced light becomes comparable with a candle or a carcel lamp.   Diaphragms CC are arranged in the cell which contains the lens, to cut off stray light.   One of these is placed at the focus of the lens, and has a small aperture.   It is easy to see that this diaphragm will cut off all light entering from a direction other than that of the source; so effectually does it do so, that observations may be made in broad daylight on any source of light, if a dark screen be placed behind it. The long box E E, Fig. 7, of about 7 feet length, is lined with black velvet,—the old-fashioned dull velvet, not that now sold with a finish, which reflects a great deal of the light incident at a certain angle. This box serves as a dark chamber, in which the intensity of the image formed by the lens is compared with a stand-

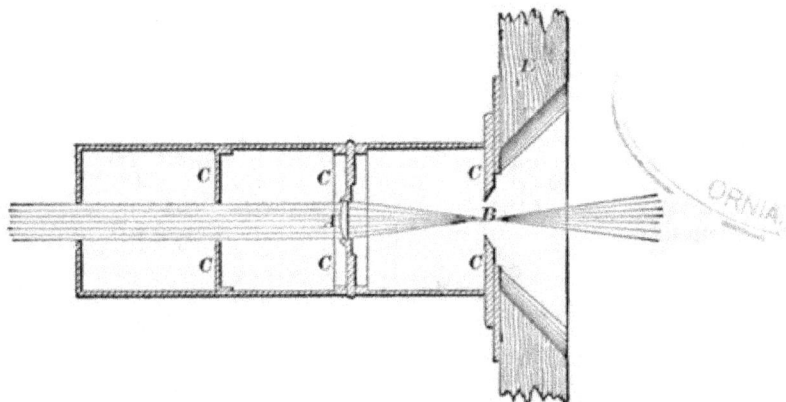

FIG. 9.—FIELD PIECE.

ard light, by means of an ordinary Bunsen's photometer F, sliding on a graduated bar.

Mr. Dallmeyer kindly had the lens made for the author: he can therefore rely upon the accuracy of its curvature and thickness; it is plano-convex, the convex side being towards the source of light. The curvature is exactly 1 inch radius and the thickness is 0.04 inch; it is made of Chance's hard crown glass, of which the refractive index for the D line in the spectrum is 1.517. The focal length f is therefore 1.933 inches.

Let u denote the distance of the source of light from

the curved surface of the lens, and v the distance of the image B of the source from the posterior focal plane. Neglecting for the moment loss by reflection at the surface of the glass, the intensity of the source is reduced by the factor $\left(\dfrac{v}{u}\right)^2$. But $\dfrac{1}{v} + \dfrac{1}{u} = \dfrac{1}{f}$, or $v = \dfrac{u f}{u - f}$; hence the factor of reduction is $\left(\dfrac{f}{u - f}\right)^2$. The effect of absorption in so small a thickness of very pure glass may be neglected; but the reflection at the surfaces will cause a loss of 8.3 per cent., which must be allowed for. This percentage is calculated from Fresnel's formulæ, which are certainly accurate for glasses of moderate refrangibility, and for moderate angles of incidence.

Suppose, for example, it is required to measure a light of 8,000 candles; if it be placed at a distance of 40 inches, it will be reduced in the ratio 467 to 1, and becomes a conveniently measurable quantity. By transmitting through colored glasses both the light from an electric lamp and that from the standard, a rough comparison may be made of the red or green in the electric light with the red or green in the standard.

A dispersive photometer, in which a lens is used in a somewhat similar manner, is described in Stevenson's "Lighthouse Illumination;" but in that case the lens is not used in combination with a Bunsen photometer, nor with any standard light. Messrs. Ayrton and Perry described a dispersive photometer with a concave lens at the meeting of the Physical Society on December 13, 1879 (*Proceedings Physical Society*, vol. III., p. 184). The convex lens possesses, however, an obvious advantage in having a real focus, at which a diaphragm to cut off stray light may be placed.

Efficiency of the Electric Arc.—To define the electrical condition of an electric arc, two quantities must be stated —the current passing, and the difference of electric potential at the ends of the two carbons. Instead of either one of these, we may, if we please, state the ratio $\frac{\text{difference of potential}}{\text{current}}$, and call it the resistance of the arc, that is to say, the resistance which would replace the arc without changing the current. But such a use of the term electric resistance is unscientific; for Ohm's law, on which the definition of electric resistance rests, is quite untrue of the electric arc; while on the other hand, for a given material of the electrodes, a given distance between them, and a given atmospheric pressure, the difference of potential on the two sides of the arc is approximately constant. The product of the difference of potential and the current is of course equal to the work developed in the arc; and this, divided by the work expended in driving the machine, may be considered as the efficiency of the whole combination. It is a very easy matter to measure these quantities. The difference of potential on the two sides of the arc may be measured by the method given by the author in his previous paper, or by an electrometer, or in other ways. The current may be measured by an Obach galvanometer, or by a suitable electro-dynamometer, or best of all, in the author's opinion, by passing the whole current, on its way to the arc, through a very small known resistance, which may be regarded as a shunt for a galvanometer of very high resistance, or to the circuit of which a very high resistance has been added.

It appears that with the ordinary carbons, and at ordinary atmospheric pressure, no arc can exist with a less difference of potential than about 20 volts; and that in ordinary work, with an arc about ¼ inch long, the difference of potential is from 30 to 50 volts. Assuming the former result, about 20 volts, for the difference of potential, the use of the curve of electromotive forces may be illustrated by determining the lowest speed at which a given machine can run and yet be capable of producing a short arc. Taking O as the origin of co-ordinates, Fig. 10, set off upon the axis of ordinates the distance $O A$

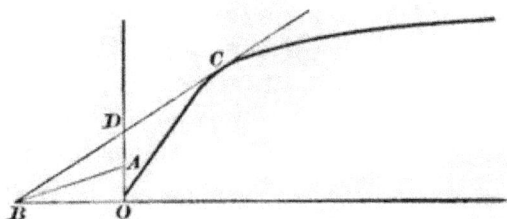

FIG. 10.

equal to 20 volts ; draw $A B$ to intersect at B the negative prolongation of the axis of abscissæ, so that the ratio $\dfrac{OA}{OB}$ may represent the necessary metallic resistance of the circuit. Through the point B, thus obtained, draw a tangent to the curve, touching it at C, and cutting $O A$ in D. Then the speed of the machine, corresponding to the particular curve employed, must be diminished in the ratio

$\dfrac{OD}{OA}$, in order that an exceedingly small arc may be just possible.

The curve may also be employed to put into a somewhat different form the explanation given by Dr. Siemens, at the Royal Society, respecting the occasional instability of the electric light as produced by ordinary dynamo-electric machines. The operation of all ordinary regulators is to part the carbons when the current is greater than a certain amount, and to close them when it is less; initially the carbons are in contact. Through the origin O, Fig. 11, draw the straight line OA, inclined at the angle repre-

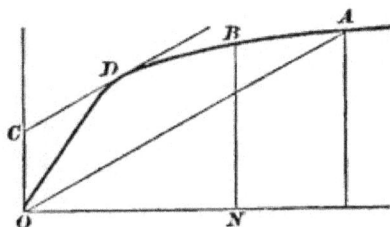

Fig. 11.

senting the resistances of the circuit other than the arc, and meeting the curve at A. The abscissa of the point A represents the current which will pass if the lamp be prevented from operating. Let $O N$ represent the current to which the lamp is adjusted; then if the abscissa of A be greater than $O N$, the carbons will part. Through N draw the ordinate $B N$, meeting the curve in the point B; and parallel to $O A$ draw a tangent $C D$, touching the curve at D. If the point B is to the right of D, or further from the origin, the arc will persist; but if B is to the left of D, or nearer to the origin, the carbons will go on parting, till the current suddenly fails and the light goes out. If B, although to the right of D, is very near to it, a very small reduction in the speed of the machine will suffice to extinguish the light. Dr. Siemens gives greater stability to the light by exciting the electromagnets of the machine by a shunt circuit, instead of by the whole current.

The success of burning more than one regulating lamp in series depends on the use in the regulator of an electromagnet excited by a high resistance wire connecting the two opposed carbons. The force of this magnet will depend upon the difference of potential in the arc, instead of depending, as in the ordinary lamp, upon the current passing. Such a shunt magnet has been employed in a variety of ways. The author has arranged it as an attachment to an ordinary regulator; the shunt magnet actuates a key, which short circuits the magnet of the lamp when the carbons are too far parted, and so causes them to close.

In conclusion the author ventures to remind engineers of the following rule for determining the efficiency of any system of electric lighting in which the electric arc is used, the arc being neither exceptionally long nor exceptionally short: Measure the difference of potential of the arc, and also the current passing through it, in volts and webers respectively; then the product of these quantities, divided * by 746, is the horse power developed in that arc.

---

* With respect to the factor 746, given above, the product of difference of potential and current was power, which could of course be given as so many foot-pounds per minute; but the number that was got by multiplying webers and volts together did not give the power in foot pounds, and it required a factor to reduce

It is then known that the difference between the horse power developed in the arc and the horse power expended to drive the machine must be absolutely wasted, and has been expended in heating either the iron of the machine or the copper conducting wires.

---

the one to the oth r, just as it required a factor to reduce gramme-centimetres, or any other measure of power, to foot-pounds. The factor in this case happened to be 746, as would be seen by referring to Everett, "Units and Physical Constants." The product of a weber and a volt was 10^7 ergs per second (p. 138), while a horse power was $7.46 \times 10^9 = 746 \times 10^7$ ergs per second (p. 25); hence the rule given.

SOME POINTS IN ELECTRIC LIGHTING.

ARTIFICIAL light is generally produced by raising some body to a high temperature. If the temperature of a body be greater than that of surrounding bodies it parts with some of its energy in the form of radiation. While the temperature is low these radiations are not of a kind to which the eye is sensitive; they are exclusively radiations less refrangible and of greater wave length than red light, and may be called infra red. As the temperature is increased the infra red radiations increase, but presently there are added radiations which the eye perceives as red light. As the temperature is further increased, the red light increases, and yellow, green and blue rays are successively thrown off in addition. On pushing the temperature to a still higher point, radiations of a wave length shorter even than violet light are produced, to which the eye is insensitive, but which act strongly on certain chemical substances; these may be called ultra violet rays. It is thus seen that a very hot body in general throws out rays of various wave lengths,—our eyes, it so happens, being only sensitive to certain of these, viz., those not very long and not very short,—and that the hotter the body the more of every kind of radiation will it throw out; but the proportion of short waves to long waves becomes vastly greater as the temperature is increased. The problem of the artificial

production of light with economy of energy is the same as
that of raising some body to such a temperature that it
shall give as large a proportion as possible of those rays
which the eye happens to be capable of feeling. For practical purposes this temperature is the highest temperature
we can produce. Owing to the high temperature at
which it remains solid, and to its great emissive power, the
radiant body used for artificial illumination is nearly always
some form of carbon. In the electric current we have an
agent whereby we can convert more energy of other forms
into heat in a small space than in any other way; and
fortunately carbon is a conductor of electricity as well as a
very refractory substance.

The science of lighting by electricity very naturally
divides itself into two principal parts—the methods of
production of electric currents, and of conversion of the
energy of those currents into heat at such a temperature
as to be given off in radiations to which our eyes are sensible. There are other subordinate branches of the subject,
such as the consideration of the conductors through which
the electric energy is transmitted, and the measurement of
the quantity of electricity passing and its potential or electric pressure. Although I shall have a word or two to say on
the other branches of the subject, I propose to occupy most
of the time at my disposal this evening with certain points
concerning the conversion of mechanical energy into electrical energy. We know nothing as to what electricity is,
and its appeals to our senses are in general less direct than
those of the mechanical phenomena of matter. The laws,
however, which we know to connect together those phenomena which we call electrical are essentially mechanical

in form, are closely correlated with mechanical laws, and may be most aptly illustrated by mechanical analogues. For example, the terms "potential," "current" and "resistance," with which we are becoming familiar in electricity, have close analogues respectively in "head," "rate of flow" and "coefficient of friction" in the hydraulic transmission of power.   Exactly as in hydraulics head multiplied by velocity of flow is power measured in foot-pounds per second or in horse power, so potential multiplied by current is power and is measurable in the same units.   The horse power not being a convenient electrical unit, Dr. Siemens has suggested that the electrical unit of power or volt-ampère should be called a watt: 746 watts are equal to one horse power. Again, just as water flowing in a pipe has inertia and requires an expenditure of work to set it in motion, and is capable of producing disruptive effects if its motion is too suddenly arrested,—as, for example, when a plug tap is suddenly closed in a pipe through which water is flowing rapidly,—so a current of electricity in a wire has inertia; to set it moving electromotive force must work for a finite time, and if we attempt to arrest it suddenly by breaking the circuit, the electricity forces its way across the interval as a spark. Corresponding to mass and moments of inertia in mechanics we have in electricity coefficients of self induction. We will now show that an electric circuit behaves as though it had inertia.   The ap-

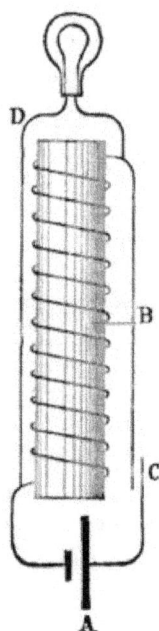

FIG. 12.

paratus we shall use is shown diagrammatically in Fig. 12.
A current from a Sellon battery A circulates round an
electromagnet B; it can be made and broken at pleasure
at C. Connected to the two extremities of the wire on the

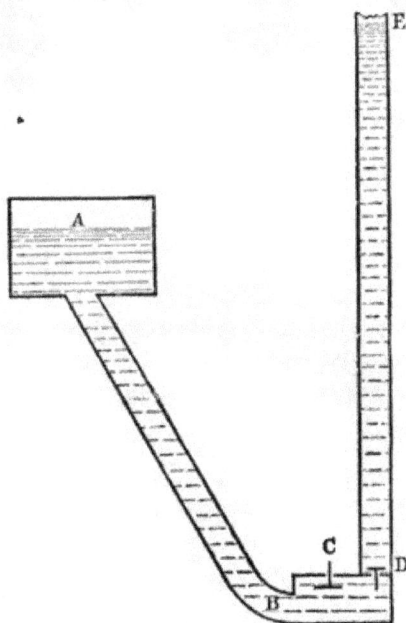

Fig. 13.

magnet is a small incandescent lamp D, lent to me by Mr.
Crompton, of many times the resistance of the coil. On
breaking the circuit, the current in the coil, in virtue of its
momentum, forces its way through the lamp, and renders
it momentarily incandescent, although all connection with
the battery, which in any case would be too feeble to send
sufficient current through the lamp, has ceased. Let us.
try the experiment, make contact, break contact. You

observe the lamp lights up. Compare with the diagram
(Fig. 13) of the hydraulic analogue, the hydraulic ram.
There a current of water suddenly arrested forces a way
for a portion of its quantity to a greater height than that
from which it fell. *A B* corresponds to the electromagnet, the valve *C* to the contact breaker, and *D E* to the
lamp. There is, however, this difference between the inertia of water in a pipe and the inertia of an electric current: the inertia of the water is confined to the water,
whereas the inertia of the electric current resides in the
surrounding medium. Hence arise the phenomena of induction of currents upon currents, and of magnets upon
moving conductors—phenomena which have no immediate
analogues in hydraulics. There is thus little difficulty to
any one accustomed to the laws of rational mechanics in
adapting the expression of those laws to fit electrical
phenomena; indeed we may go so far as to say that the
part of electrical science with which we have to deal this
evening is essentially a branch of mechanics, and as such
I shall endeavor to treat it.

This is neither the time nor the place for setting forth
the fundamental laws of electricity, but I cannot forbear
from showing you a mechanical illustration, or set of
mechanical illustrations, of the laws of electrical induction,
first discovered by Faraday. I have here a model, Fig. 14,
which was made to the instructions of the late Professor
Clerk Maxwell, to illustrate the laws of induction. It
consists of a pulley *P*, which I now turn with my hand,
and which represents one electric circuit, its motion the
current therein. Here is a second pulley, *S*, representing
a second electric circuit. These two pulleys are geared

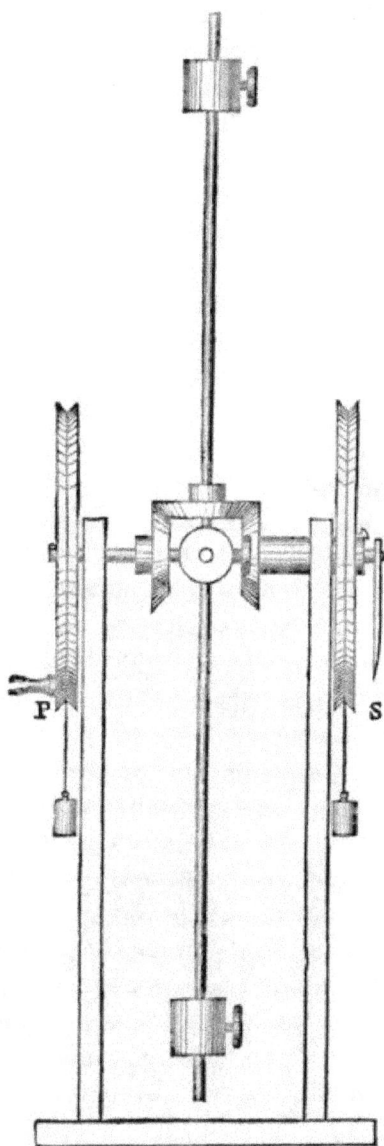

FIG. 14.

together by a simple differential train, such as is sometimes used for a dynamometer. The intermediate wheel of the train, however, is attached to a balanced flywheel, the moment of inertia of which can be varied by moving inwards or outwards these four brass weights. The resistances of the two electric circuits are represented by the friction on the pulleys of two strings, the tension of which can be varied by tightening these elastic bands. The differential train, with its flywheel, represents the medium, whatever it may be, between the two electric conductors. The mechanical properties of this model are of course obvious enough. Although the mathematical equations which represent the relation between one electric conductor and another in its neighborhood are the same in form as the mathematical equations which represent the mechanical connection between these two pulleys, it must not be assumed that the magnetic mechanism is completely represented by the model. We shall now see how the model illustrates the action of one electric circuit upon another. You know that Faraday discovered that if you have two closed conductors arranged near to and parallel to each other, and if you cause a current of electricity to begin to flow in the first, there will arise a temporary current in the opposite direction in the second. This pulley, marked P on the diagram, represents the primary circuit, and the pulley marked S on the diagram the secondary circuit. We cause a current to begin to flow in the primary, or turn the pulley P; an opposite current is induced in the secondary circuit, or the pulley S turns in the opposite direction to that in which we began to move the pulley P. The effect is only temporary; resistance speedily stops the current in the secondary circuit, or, in the mechanical model, friction the rotation of the pulley S. I now gradually stop the motion of P; the pulley S moves in the direction in which P was previously moving, just as Faraday found that the cessation of the primary current induced in the secondary circuit a current in the same direction as that which had existed in the primary. If there were a large number of convolutions or coils in the secondary circuit, but that circuit were not completed, but had an air space interrupting its continuity, an experiment with the well known Ruhmkorff coil would show you that when the current was *suddenly* made to cease to flow in the primary circuit, so great 'an electromotive force would be exerted in the secondary circuit that the electricity would leap across the space as a spark. I will now show you what corresponds to a spark with this mechanical model. The secondary pulley S shall be held by passing a thread several times round it. I gradually produce the current in the primary circuit. I will now suddenly stop this primary current: you observe that the electromotive force is sufficient to break the thread. The inductive effects of one electric circuit upon another depend not alone on the dimensions and form of the two circuits, but on the nature of the material between them. For example, if we had two parallel circular coils, their inductive effects would be very considerably enhanced by introducing a bar of iron in their common axis. We can imitate this effect by moving outwards or inwards these brass weights. In the experiment I have shown you the weights have been some distance from the axis in order to obtain considerable effect, just as in the Ruhmkorff coil an iron core is introduced within the primary circuit. I will now do what is equivalent to removing the core: I will bring the weights nearer to the axis, so that my flywheel shall have less moment of inertia. You observe that the inductive effects are very much less marked than they were before. With the same electromagnet which we used before, but differ-

ently arranged, we will show what we have just illustrated—the induction of one circuit on another. Referring to Fig. 15, coil _A B_ corresponds to wheel _P; C D_ to wheel _S_, and the iron core to the fly-wheel and differential gear. The resist-ance of a lamp takes the place of the friction of the string on _S_. As we make and break the circuit you see the effect of the induced current in rendering the lamp incandescent. So far I have been illustrating the phenomena of the induc-tion of one current upon another. I will now show on the model that a current in a single electric circuit has momen-tum. The secondary wheel shall be firmly held; it shall have no conductivity at all—that is, its electrical effect shall be as though it were not there. I now

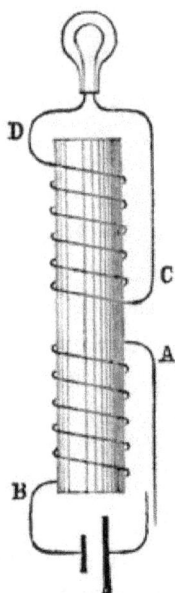

FIG. 15.

cause a current to begin to flow in the primary circuit, and it is obvious enough that a certain amount of work must be done to bring it up to a certain speed. The an-gular velocity of the flywheel is half that of the pulley representing the primary circuit. Now suppose that the two pulleys were connected together in such a way that

they must have the same angular velocity in the same
direction.   This represents the coil having twice as many
convolutions as it had before.   A little
consideration will show that I must do
four times as much work to give the
primary pulley the same velocity that
it attained before; that is to say, that
the coefficient of self induction of a coil
of wire is proportional to the square of
the number of convolutions.   Again,
suppose that these two wheels were so
geared together that they must always
have equal and opposite velocities, you
can see that a very small amount of
work must be done in order to give the
primary wheel the velocity which we
gave to it before.   Such an arrangement
of the model represents an electric circuit, the coefficient of induction of which
is exceedingly small, such as the coils
that are wound for standard resistances;
the wire is there wound double, and
the current returns upon itself, as shown
in Fig. 16.

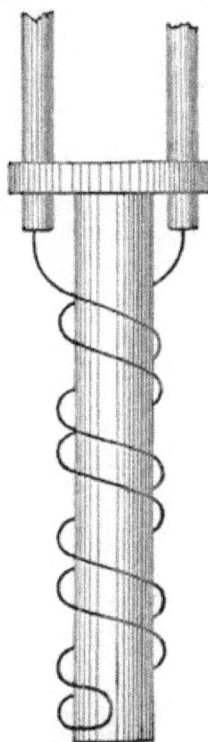

FIG. 16.

In the widest sense, the dynamo-electric machine may
be defined as an apparatus for converting mechanical
energy into the energy of electrostatic charge, or mechanical power into its equivalent electric current through a
conductor.   Under this definition would be included the
electrophorus and all frictional machines; but the term is
used, in a more restricted sense, for those machines which

produce electric currents by the motion of conductors in a
magnetic field, or by the motion of a magnetic field in the
neighborhood of a conductor. The laws on which the
action of such machines is based have been the subject of
a series of discoveries. Oersted discovered that an electric
current in a conductor exerted force upon a magnet;
Ampère that two conductors conveying currents generally
exerted a mechanical force upon each other. Faraday discovered—what Helmholtz and Thomson subsequently
proved to be the necessary consequence of the mechanical
reactions between conductors conveying currents and magnets—that if a closed conductor move in a magnetic field,
there will be a current induced in that conductor in one
direction if the number of lines of magnetic force passing
through the conductor was increased by the movement; in
the other direction if diminished. Now all dynamo-electric
machines are based upon Faraday's discovery. Not only
so; but however elaborate we may wish to make the analysis
of the action of a dynamo machine, Faraday's way of pre-
senting the phenomena of electromagnetism to the mind
is in general our best point of departure. The dynamo
machine, then, essentially consists of a conductor made to
move in a magnetic field. This conductor, with the external circuit, forms a closed circuit in which electric currents
are induced as the number of lines of magnetic force passing through the closed circuit varies. Since, then, if the
current in a closed circuit be in one direction when the
number of lines of force is increasing, and in the opposite
direction when they are diminishing, it is clear that the
current in each part of such circuit which passes through
the magnetic field must be alternating in direction, unless,

indeed, the circuit be such that it is continually cutting more and more lines of force, always in the same direction. Since the current in the wire of the machine is alternating, so also must be the current outside the machine, unless something in the nature of a commutator be employed to reverse the connections of the internal wires in which the current is induced, and of the external circuit. We have, then, broadly, two classes of dynamo-electric machines— the simplest, the alternating current machine, where no commutator is used; and the continuous current machine, in which a commutator is used to change the connection of the external circuit just at the moment when the direction of the current would change. The mathematical theory of the alternate current machine is comparatively simple. To fix ideas, I will ask you to think of the alternate current Siemens machine, which Dr. Siemens exhibited here three weeks ago. We have there a series of magnetic fields of alternate polarity, and through these fields we have coils of wire moving; these coils constitute what is called the armature; in them are induced the currents which give a useful effect outside the machine. Now I am not going to trouble you to go through the mathematical equations, simple though they are, by which the following formulæ are obtained:—

$$I = A \sin \frac{2\,\pi\,t}{T} \qquad\qquad \text{(I.)}$$

$$E = \frac{2\,\pi\,A}{T} \cos \frac{2\,\pi\,t}{T} \qquad\qquad \text{(II.)}$$

$$x = \frac{2\pi A}{T} \; \frac{\cos 2\pi \dfrac{t-\tau}{T}}{\sqrt{\left(\dfrac{2\pi\gamma}{T}\right)^2 + R^2}} \qquad\qquad \text{(III.)}$$

$$\text{Tan}\; \frac{2\pi\tau}{T} = \frac{2\pi\gamma}{RT} \qquad\qquad \text{(IV.)}$$

$$\Theta R \frac{2\pi^2 A^2}{T^2} \; \frac{I}{\left(\dfrac{2\pi\gamma}{T}\right)^2 + R^2} \qquad\qquad \text{(V.)}$$

$$R = \frac{2\pi\gamma}{T} \qquad\qquad \text{(VI.)}$$

T represents the periodic time of the machine; that is, in the case of a Siemens machine having eight magnets on each side of the armature, T represents the time of one-fourth of a revolution. I represents the number of lines of force embraced by the coils of the armature at the time t. I must be a periodic function of t, in the simplest form represented by Equation I. Equation II. gives E the electromotive force acting at time t upon the circuit. Having given the electromotive force acting at any time, it would appear at first sight that we had nothing to do but to divide that electromotive force by the resistance R of the whole circuit, to obtain the current flowing at that time. But if we were to do so we should be landed in error, for the conducting circuit has other properties besides resistance. I pointed out to you that it had a property of momentum represented by its coefficient of self induction, called γ in the formula; and when we are dealing with rapid changes of current, it plays as important a part as the resistance. Formula III. gives the current x, flowing at any time, and you will observe that it shows two things: first, the maximum current is less than it would be if there were no self induction; secondly, it attains its maximum at a later time. This retardation is represented by the letter τ, and its amount is determined by Formula **IV.** At a given speed of rotation, the amount of **electrical work developed in** the machine in any **time** Θ **is given by Formula V.** It is greatest when $R = \dfrac{2\,\pi\,\gamma}{T}$. **From these for-** mulæ we see that the current is diminished either by increasing γ or increasing R; also the moment of reversal of current is not coincident with the moment of reversal of electromotive force, but occurs later, by an amount depending on the relative **magnitudes of** γ and R. They show us that although by doubling the velocity of the machine we really double the electromotive force at any time, we **do** not double the **current** passing, nor the **work done by the** machine; but we may see that if we double **the velocity** of the machine we may work through double the **external** resistance and still obtain the same **current. In what pre-** cedes, it has been assumed that the copper **wires are** the only conducting bodies moving in the magnetic field. **In** many cases **the moving** wire coils of these machines have iron cores, **the iron** being in some cases **solid, in** others more or less divided. **It** is found that if such **machines** are run on open circuit, **that is, so** that **no current circu-** lates in the armatures, the iron **becomes hot,** very much hotter than when the circuit of the copper wire is **closed,**

In some cases this phenomenon is so marked that the machine actually takes more to drive it, when the machine is on open circuit, than when it is short circuited. The explanation is that on open circuit currents are induced in the iron cores, but that when the copper coils are closed the current in them diminishes by induction the current in the iron. The effect of currents in the iron cores is not alone to waste energy and heat the machine; but for a given intensity of field and speed of revolution the external current produced is diminished. The cure of the evil is to subdivide the moving iron as much as possible, in directions perpendicular to those in which the current tends to circulate.

There remains one point of great practical interest in connection with alternate current machines: How will they behave when two or more are coupled together to aid each other in doing the same work? With galvanic batteries we know very well how to couple them, either in parallel circuit or in series, so that they shall aid, and not oppose, the effects of each other; but with alternate current machines, independently driven, it is not quite obvious what the result will be, for the polarity of each machine is constantly changing. Will two machines, coupled together, run independently of each other, or will one control the movement of the other in such wise that they settle down to conspire to produce the same effect, or will it be into mutual opposition? It is obvious that a great deal turns upon the answer to this question, for in the general distribution of electric light it will be desirable to be able to supply the system of conductors from which the consumers draw by separate machines,

which can be thrown in and out at pleasure.   Now I know
it is a common impression that alternate current machines
cannot be worked together, and that it is almost a necessity
to have one enormous machine to supply all the consumers
drawing from one system of conductors.   Let us see how
the matter stands.   Consider two machines independently
driven, so as to have approximately the same periodic time

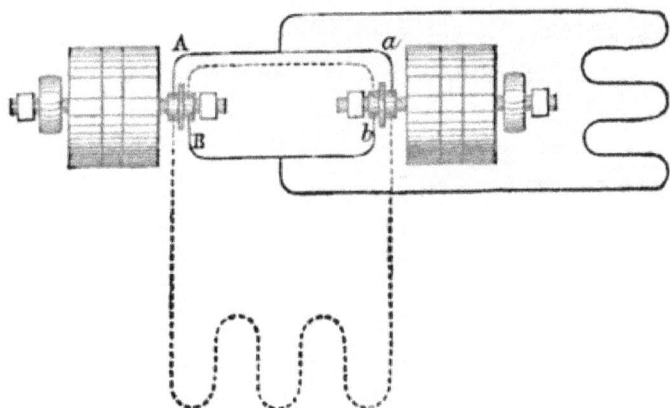

Fig. 17.

and the same electromotive force.   If these two machines
are to be worked together, they may be connected in one
of two ways: they may be in parallel circuit with regard to
the external conductor, as shown by the full line in Fig. 17,
that is, their currents may be added algebraically and sent
to the external circuit, or they may be coupled in series, as
shown by the dotted line, that is, the whole current may
pass successively through the two machines, and the
electromotive force of the two machines may be added,

instead of their currents. The latter case is simpler. Let us consider it first. I am going to show that if you couple two such alternate current machines in series they will so control each other's phase as to nullify each other, and that you will get no effect from them; and, as a corollary from that, I am going to show that if you couple them in parallel circuit they will work perfectly well together, and the currents they produce will be added; in fact, that you cannot drive alternate current machines tandem, but that you may drive them as a pair, or, indeed, any number abreast. In diagram, Fig. 18, the horizontal line of ab-

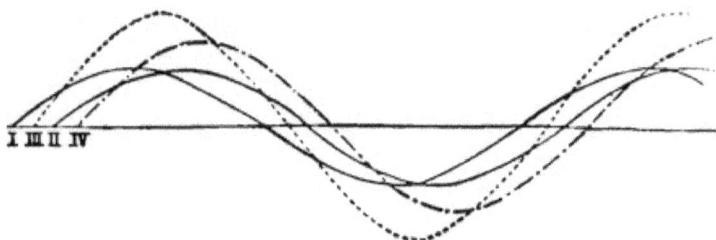

Fig. 18.

scissæ represents the time advancing from left to right; the full curves represent the electromotive forces of the two machines not supposed to be in the same phase. We want to see whether they will tend to get into the same phase or to get into opposite phases. Now, if the machines are coupled in series, the resultant electromotive force on the circuit will be the sum of the electromotive forces of the two machines. This resultant electromotive force is represented by the broken curve *III*. By what we have already seen in Formula IV., the phase of the current must lag behind the phase of the electromotive force, as is shown in the diagram by curve IV, thus ———. ———. ———. Now the work done in any machine is represented by the sum of the products of the currents and of the electromotive forces, and it is clear that, as the phase of the current is more near to the phase of the lagging machine II than to that of the leading machine I, the lagging machine must do more work in producing electricity than the leading machine; consequently its velocity will be retarded, and its retardation will go on until the two machines settle down into exactly opposite phases, when no current will pass. The moral, therefore, is, do not attempt to couple two independently driven alternate current machines in series. Now for the corollary: A, B, Fig. 17, represent the two terminals of an alternate current machine; a, b the two terminals of another machine independently driven. A and a are connected together, and B and b. So regarded, the two machines are in series, and we have just proved that they will exactly oppose each other's effects, that is, when A is positive, a will be positive also; when A is negative, a is also negative. Now, connecting A and a through the comparatively high resistance of the external circuit with B and b, the current passing through that circuit will not much disturb, if at all, the relations of the two machines. Hence, when A is positive, a will be positive, and when A is negative, a will be negative also; precisely the condition required that the two machines may work together to send a current into the external circuit. You may, therefore, with confidence, attempt to run alternate current machines in parallel circuit for the purpose of producing any external effect. I might easily show that the same applies to a larger number; hence there is no more difficulty in feeding a system of conductors from a number of alternate current machines than there is in feeding it from a number of continuous current machines. A little care is only required that the machine shall be thrown in when it has attained something like its proper velocity. A further corollary is that alternate currents with alternate current machines as motors may theoretically be used for the transmission of power.*

It is easy to see that, by introducing a commutator revolving with the armature, in an alternate current machine, and so arranged as to reverse the connection between the armature and the external circuit just at the time when the current would reverse, it is possible to obtain a current constant always in direction; but such a current would be far from constant in intensity, and would certainly not accomplish all the results that are obtained in modern continuous current machines. This irregularity may, however, be reduced to any extent by multiplying the wires of the armature, giving each its own connection to the outer circuit, and so placing them that the electromotive force attains a maximum successively in the several coils. A practically uniform electric current was first commercially produced with the ring armature of Pacinotti, as perfected by Gramme. The Gramme machine is represented diagrammatically in Fig. 19. The armature consists

---

* Of course in applying these conclusions it is necessary to remember that the machines only *tend* to control each other, and that the control of the motive power may be predominant, and *compel* the two or more machines to run at different speeds.

of an anchor ring of iron wire, the strands more or less insulated from each other. Round this anchor ring is wound a continuous endless coil of copper wire; the armature moves in a magnetic field, produced by permanent or electromagnets with diametrically opposite poles, marked N and S. The lines of magnetic force may be regarded as passing into the ring from N, dividing, passing round the ring and across to S. Thus the coils of wire, both near to N and near to S, are cutting through a very strong magnetic field; consequently there will be an intense inductive action. The inductive action of the coils near N being equal and opposite to the inductive action of the coils near S, it results that there will be strong positive and negative electric potential at the extremities of a diameter perpendicular to the line NS. The electromotive force produced is made use of to produce a

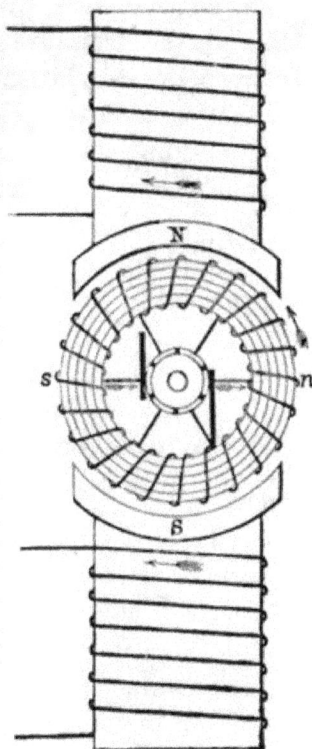

Fig. 19.

current external to the machine; thus the endless coil of the armature is divided into any number of sections, in the diagram into six for convenience, usually into sixty or eighty, and the junction of each pair of sections is connected by a wire to a plate of the commutator fixed upon the shaft which carries the armature; collecting brushes make contact with the commutator, as shown in the diagram. If the external resistance were enormously high, so that very little current, or none at all, passed through the armature, the greatest difference of potential between the two brushes would be found when they made contact at points at right angles to the line between the magnets; but when a current passes in the armature, this current causes a disturbing effect upon the magnetic field. Every time the contact of the brushes changes from one contact plate to the next, the current in a section of the copper coil is reversed, and this reversal has an inductive effect upon all the other coils of the armature. You may take it from me that the net result on any one coil is approximately the same as if that coil alone were moved, and all the other coils were fixed, and there were no reversals of current in them. Now you can easily see that the magnetic effect of the current circulating in the coils of the armature will be to produce a north pole at n and a south pole at s. This will displace the magnetic field in the direction of rotation. If, then, we were to keep the contact points the same as when no current was passing, we should short circuit the sections of the armature at a time when they were cutting through the lines of magnetic force, with a result that there would be vigorous sparks between the collecting brushes and the commutator. To avoid this, the brushes must follow the magnetic field, and also be displaced in the direction of rotation, this displacement being greater as the current in the armature is greater in proportion to the magnetic field. The net effect of this disturbing effect of the current in

the armature reacting upon itself is, then, to displace the
neutral points upon the commutator, and consequently
somewhat to diminish the effective electromotive force.
It is best to adjust the brushes to make contact at a point
such that, with the current then passing, flashing is reduced to a minimum; but this point does not necessarily
coincide with the point which gives maximum difference
of potential. The magnetic field in the Gramme and
other continuous dynamo-electric machines may be produced in several ways. Permanent magnets of steel may be
used, as in some of the smaller machines now made, and in
all the earlier machines; these are frequently called magneto machines. Electromagnets excited by a current
from a small dynamo-electric machine were introduced
by Wilde; these may be described shortly as dynamos
with separate exciters. The plan of using the whole
current from the armature of the machine itself, for
exciting the magnets, was proposed almost simultaneously
by Siemens, Wheatstone, and S. A. Varley. A dynamo
so excited is now called a series dynamo. Another method
is to divide the current from the armature, sending the
greater part into the external circuit, and a smaller portion through the electromagnet, which is then of very
much higher resistance. Such an arrangement is called a
shunt dynamo. A combination of the last two methods
has been recently introduced, for the purpose of maintaining constant potential. The magnet is partly excited by a circuit of high resistance, a shunt to the
external circuit, and partly by coils conveying the
whole current from the armature. All but the first
two arrangements named depend on residual magnetism

to initiate the current, and below a certain speed of rotation give no practically useful electromotive force. A dynamo machine is, of course, not a perfect instrument for converting mechanical energy into the energy of electric current. Certain losses inevitably occur. There is, of course, the loss due to friction of bearings, and of the collecting brushes upon the commutator; there is also the loss due to the production of electric currents in the iron of the machine. When these are accounted for, we have the actual electrical effect of the machine in the conducting wire; but all of this is not available for external work. The current has to circulate through the armature, which inevitably has electrical resistance; electrical energy must, therefore, be converted into heat in the armature of the machine. Energy must also be expended in the wire of the electromagnet which produces the field, for the resistance of this also cannot be reduced beyond a certain limit. The loss by the resistance of the wires of the armature and of the magnets greatly depends on the dimensions of the machine. About this I shall have to say a word or two presently. To know the properties of any machine thoroughly, it is not enough to know its efficiency and the amount of work it is capable of doing: we need to know what it will do under all circumstances of varying resistance or varying electromotive force. We must know, under any given conditions, what will be the electromotive force of the armature. Now this electromotive force depends on the intensity of the magnetic field, and the intensity of the magnetic field depends on the current passing round the electromagnet and the current in the armature. The current, then, in the machine is the proper independent variable in terms of which to

express the electromotive force. The simplest case is that
of the series dynamo, in which the current in the electromagnet and in the armature is the same, for then we have
only one independent variable. The relation between the
electromotive force and current is represented by such a
curve as is shown in the diagram, Fig. 20. The abscissæ,

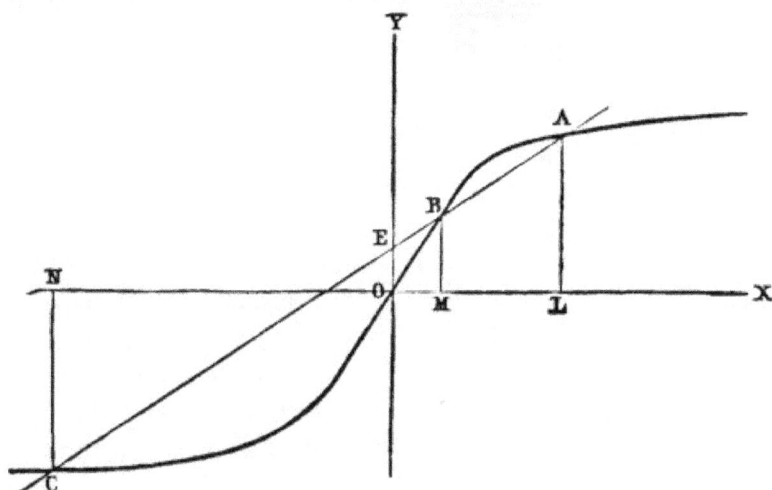

Fig. 20.

measured along $O X$, represent the current, and the ordinates represent the electromotive force in the armature.
When four years ago I first used this curve, for the purpose of expressing the results of my experiments on the
Siemens dynamo machine, I pointed out that it was capable
of solving almost any problem relating to a particular
machine, and that it was also capable of giving good indications of the results of changes in the winding of the
magnets or of the armatures of such machines. Since then
M. Marcel Deprez has happily named such curves " char-

acteristic curves." I will give you one or two illustrations
of their use. A complete characteristic of a series dynamo
does not terminate at the origin, but has a negative branch,
as shown in the diagram; for it is clear that by reversing
the current through the whole machine the electromotive
force is also reversed. Suppose a series dynamo is used for
charging an accumulator, and is driven at a given speed,
what current will pass through it? The problem is easily
solved. Along $O\,Y$, Fig. 20, set off $O\,E$ to represent the
electromotive force of the accumulator, and through E draw
the line $C\,E\,B\,A$, making an angle with $O\,X$, such that its
tangent is equal to the resistance of the whole circuit, and
cutting the characteristic curve, as it in general will do, in
three points, A, B, and C. We have, then, three answers to
the question. The current passing through the dynamo
will be either $O\,L$, $O\,M$, or $O\,N$, the abscissæ of the points
where the line cuts the curve. $O\,L$ represents the current
when the dynamo is actually charging the accumulator.
$O\,M$ represents a current which could exist for an instant,
but which would be unstable, for the least variation would
tend to increase. $O\,N$ is the current which passes if the
current in the dynamo should get reversed, as it is very
apt to do when used for this purpose. The next illustration
is rather outside my subject, but shows another method of
using the characteristic curve. Many of you have heard of
Jacobi's law of maximum effect of transmitting work by
dynamo machines. It is this: Supposing that the two
dynamo machines were perfect instruments for converting
mechanical energy into electrical energy, and that the generating machine were run at constant velocity, while the
receiving machine had a variable velocity, the greatest

amount of work would be developed in the receiving
machine when its electromotive force was one-half that of
the generating machine; then the efficiency would be one-
half, and the electrical work done by the generating machine
would be just one-half of what it would be if the receiving
machine were forcibly held at rest. Now this law is strictly
true if, and only if, the electromotive force of the generating machine is independent of the current passing through

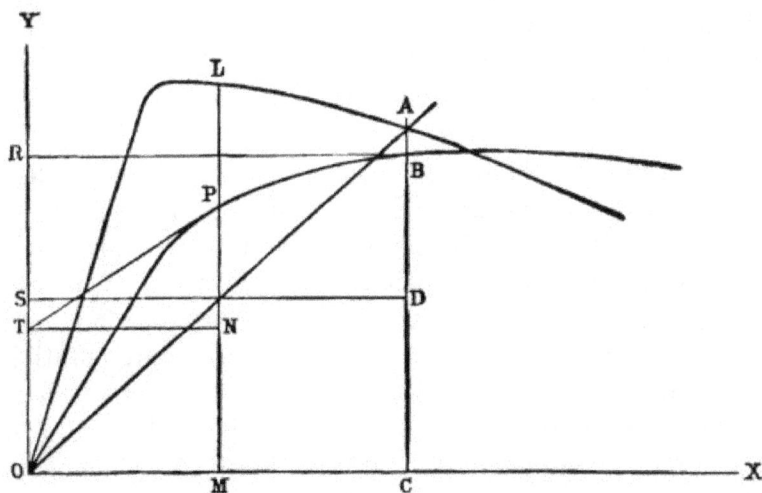

Fig. 21.

its armature. What I am now going to do is to give you a
construction for determining the maximum work which
can be transmitted when the electromotive force of the
generating machine depends on the current passing through
the armature, as, indeed, it nearly always does. Referring to
Fig. 21, $O\,PB$ is the characteristic curve of the generating
machine. Construct a derived curve thus: at successive

points P of the characteristic curve draw tangents PT; draw TN parallel to OX, cutting PM in N; produce MP to L, making LP equal PN; the point L gives the derived curve, which I want. Now, to find the maximum work which can be transmitted, draw OA at such an angle with OX that its tangent is equal to twice the resistance of the whole circuit, cutting the derived curve in A. Draw the ordinate AC, cutting the characteristic curve in B; bisect AC at D. The work expended upon the generating machine would be represented by the parallelogram $OCBR$, the work wasted in resistance by $OCDS$, and the work developed in the receiving machine by the parallelogram $SDBR$.

When the dynamo machine is not a series dynamo, but the currents in the armature and in the electromagnet, though possibly dependent upon each other, are not necessarily equal, the problem is not quite so simple. We have, then, two variables, the current in the electromagnet and the current in the armature; and the proper representation of the properties of the machine will be by a characteristic surface such as that illustrated by this model, Fig. 22. Of the three co-ordinate axes, OX represents the current in the magnet, OY represents the current in the armature, not necessarily to the same scale, and OZ the electromotive force. By the aid of such a surface as this, one may deal with any problem relating to a dynamo machine, no matter how its electromagnets and its armature are connected together. Let us apply the model to find the characteristic of a series dynamo. Take a plane through OZ, the axis of electromotive force, and making such an angle with the plane OX, OZ that its tangent is equal to

current unity on axis OY, divided by current unity on
axis OX. This plane cuts the surface in a curve. The
projection of this curve on the plane OX, OZ is the
characteristic curve of the series dynamo. This model only
shows an eighth part of the complete surface. If any of
you should interest yourselves about the other seven parts,
which are not without interest, remember that it is assumed

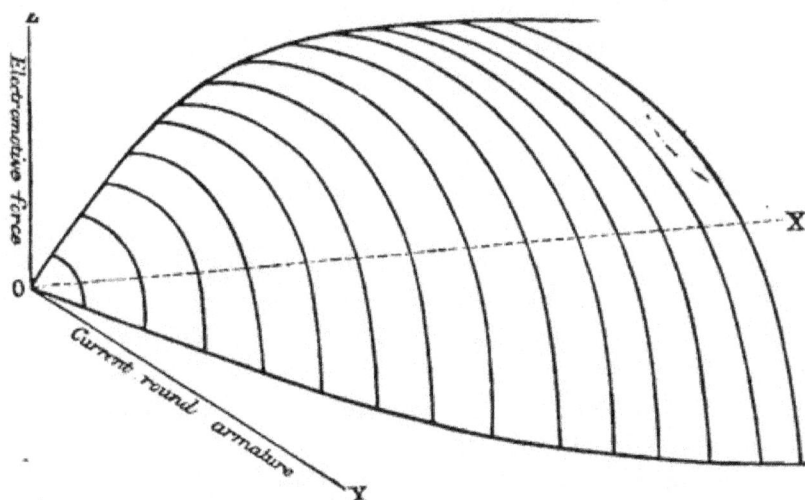

FIG. 22.

that the brushes always make contact with the commutator at the point of no flashing, if there is one. Of course
in actual practice one would not use the model of the
surface, but the projections of its sections. While I am
speaking of characteristic curves there is one point I will
just take this opportunity of mentioning. Three years ago
Mr. Shoolbred exhibited the characteristic curve of a
Gramme machine, in which, after the current attained to a

certain amount, the electromotive force began to fall. I then said that I thought there must be some mistake in the experiment. However, subsequent experiments have verified the fact; and when one considers it, it is not very difficult to see the explanation. It lies in this: after the current attains to a certain amount the iron in the machines becomes magnetically nearly saturated, and consequently an increase in the current does not produce a corresponding increase in the magnetic field. The reaction, however, between the different sections of the wire on the armature goes on increasing indefinitely, and its effect is to diminish the electromotive force.

A little while ago I said that the dimensions of the machine had a good deal to do with its efficiency. Let us see how the properties of a machine depend upon its dimensions. Suppose two machines alike in every particular excepting that the one has all its linear dimensions double those of the other; obviously enough all the surfaces in the larger would be four times the corresponding surfaces in the smaller, and the weights and volumes of the larger would be eight times the corresponding weights in the smaller machine. The electrical resistances in the larger machine would be one-half those of the smaller. The current required to produce a given intensity of magnetic field would be twice as great in the larger machine as in the smaller. In the diagram (Fig. 23) are shown the comparative characteristic curves of the two machines, when driven at the same speed. You will observe that one curve is the projection of the other, having corresponding points with abscissæ in the ratio of one to two, and the ordinates in the ratio of one to four. Now at first sight it

would seem as though, since the wire on the magnet and
armature of the larger machine has four times the section
of that of the smaller, four times the current could be
carried, that consequently the intensity of the magnetic
field would be twice as great and its area would be four
times as great, and hence the electromotive force eight
times as great; and, since the current in the armature also
is supposed to be four times as great, that the work done

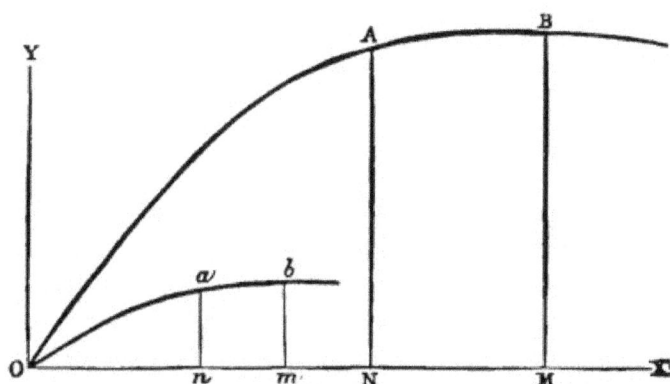

FIG. 23.

by the larger machine would be thirty-two times as much
as that which would be done by the smaller. Practically,
however, no such result can possibly be obtained, for a
whole series of reasons. First of all, the iron of the magnets becomes saturated, and consequently, instead of getting
eight times the electromotive force, we should only get four
times the electromotive force. Secondly, the current which
we can carry in the armature is limited by the rate at which
we can get rid of the heat generated in the armature. This
we may consider as proportional to its surface; consequently

we must only waste four times as much energy in the armature of the larger machine as in the smaller one, instead of eight times, as would be the case if we carried the current in proportion to the section of the wire. Again, the larger machine cannot run at so great an angular velocity as the smaller one. And lastly, since in the larger machine the current in the armature is greater in proportion to the saturated magnetic field than it is in the smaller one, the displacement of the point of contact of the brushes with the commutator will be greater. However, to cut the matter, about which one might say a great deal, short, one may say that the capacity of similar dynamo machines is pretty much proportionate to their weight, that is, to the cube of their linear dimensions; that the work wasted in producing the magnetic field will be directly as the linear dimensions; and that the work wasted in heating the wires of the armature will be as the square of the linear dimensions. Now let us see how this would practically apply. Suppose we had a small machine capable of producing an electric current of 4 h. p., that of this 4 h. p. 1 was wasted in heating the wires of the armature, and 1 in heating the wires of the magnet; 2 would be usefully applied outside. Now if we doubled the linear dimensions we should have a capacity of 32 h. p., of which 2 only, if suitably applied, would be required to produce the magnetic field, and 4 would be wasted in heating the wires of the armature, leaving 26 h. p. available for useful work outside the machine—a very different economy from that of the smaller machines. But if we again doubled the linear dimensions of our machine, we should by no means obtain a similar increase of effect. A consideration of the properties of similar machines has

another very important practical use. As you all know,
Mr. Froude was able to control the design of ironclad ships
by experiments upon models made in paraffin wax. Now
it is a very much easier thing to predict what the performance of a large dynamo machine will be, from laboratory
experiments made upon a model of a very small fraction of
its dimensions. As a proof of the practical utility of such
methods, I may say that by laboratory experiments I have
succeeded in increasing the capacity of the Edison machines
without increasing their cost, and with a small increase of
their percentage of efficiency, remarkably high as that
efficiency already was.

I might occupy your time with considerations as to the
proper proportion of conductors, and explain Sir W.
Thomson's law that the most economical size of a copper
conductor is such that the annual charge for interest and
depreciation of the copper of which it is made shall be
equal to the cost of producing the power which is wasted
by its resistance. But the remaining time will, perhaps, be
best spent in considering the production of light from the
energy of electric currents. You all know that this is done
commercially in two ways—by the electric arc, and by the
incandescent lamp; as the arc lamp preceded the incandescent lamp historically, we will examine one or two points
connected with it first.

I have here all that is necessary to illustrate the electric
arc, viz., two rods of carbon supported in line with each
other, and so mounted that they can be approached or withdrawn. Each carbon is connected with one of the poles of
the Edison dynamo machine which is supplying electricity
to the incandescent lamps which illuminate the whole of

this building. A resistance is interposed in the circuit of the lamp, because the electromotive force of the machine is much in excess of what the lamp requires. I now approach the carbons, bring them into contact, and again separate them slightly; you observe that the break does not stop the current, which forces its way across the space. I increase the distance between the carbons, and you observe the electric arc between their extremities; at last it breaks, having attained a length of about 1 inch. Now the current has hard work to cross this air space between the carbons, and the energy there developed is converted into heat, which raises the temperature of the ends of the carbon beyond any other terrestrial temperature. There are several points of interest I wish to notice in the electric arc. Both carbons burn away in the air, but there is also a transference of carbon from the positive to the negative carbon; therefore, although both waste away, the positive carbon wastes about twice as fast as the negative. With a continuous current, such as we are using now, the negative carbon becomes pointed, while the positive carbon forms a crater or hollow; it is this crater which becomes most intensely hot and radiates most of the light; hence the light is not by any means uniformly distributed in all directions, but is mainly thrown forward from the crater in the positive carbon. This peculiarity is of great advantage for some purposes, such, for example, as military or naval search lights; but it necessitates, in describing the illuminating power of an arc light, some statement of the direction in which the measurement was made. On account of its very high temperature the arc light sends forth a very large amount of visible radiation, and is therefore very economical of electrical energy. For the same reason its light contains a very large proportion of rays of high refrangibility, blue and ultra violet. I have measured the red light of an electric arc against the red of a candle, and have found it to be 4,700 times as great, and I have measured the blue of the same arc light against the blue of the same candle, and found it to be 11,380 times as great. The properties of an electric arc are not those of an ordinary conductor. Ohm's law does not apply. The electromotive force and the current do not by any means bear to each other a constant ratio. Strictly speaking, an electric arc cannot be said to have an electric resistance measurable in ohms. We will now examine the electrical properties of the arc experimentally. In the circuit with the lamp is a Thomson graded current galvanometer for measuring the current passing in ampères; connected to the two carbons is a Thomson graded potential galvanometer for measuring the difference of potential between them in volts. We have the means of varying the current by varying the resistance, which I have already told you is introduced into the circuit. We will first put in circuit the whole resistance available, and will adjust the carbons so that the distance between them is, as near as I can judge, ⅛ inch. We will afterwards increase the current and repeat the readings. The results are given in Table III.

TABLE III.

Current Galvanometer.	Potential Galvanometer.	Ampères.	Volts.	Watts.	H. P.
6.2	12.0	9.9	35	346	0.46
9.3	12.0	11.9	35	521	0.70
11.5	11.8	18.4	34	626	0.84

If the electrical properties of the arc were the same as those of a continuous conductor, the volts would be in proportion to the ampères, if correction were made for change of temperature; you observe that instead of that the potential is nearly the same in the two cases. We may say, with some approach to accuracy, that, with a given length of arc, the arc opposes to the current an electromotive force nearly constant, almost independent of the current. This was first pointed out by Edlund. If you will speak of the resistance of the electric arc, you may say that the resistance varies inversely as the current. Take the last experiment: by burning 4 cubic feet of gas per hour we should produce heat energy at about the same rate. I leave any of you to judge of the comparative illuminating effects. It is not my purpose to describe the mechanisms which have been invented for controlling the feeding of the carbons as they waste away. Several lamps lent by Messrs. Siemens Brothers—to whom I am indebted for the lamp and resistance I have just been using—lie upon the table for inspection. An electric arc can also be produced by an alternate current. Its theory may be treated mathematically, and is very interesting, but time will not allow us to go into it. I will merely point out this: there is some theoretical reason to suppose that an alternate current arc is in some measure less efficient than one produced by a continuous current. The efficiency of a source of light is greater as the mean temperature of the radiating surface is greater. The maximum temperature in an arc is limited probably by the temperature of volatilization of carbon; in an alternate current arc the current is not constant, therefore the mean temperature is less than the maximum temperature; in a continuous current arc, the current being constant, the mean and maximum temperatures are equal, therefore in a continuous current arc the mean temperature is likely to be somewhat higher than in an alternate current arc.

We will now pass to the simpler incandescent light. When a current of electricity passes through a continuous conductor, it encounters resistance, and heat is generated, as was shown by Joule, at a rate represented by the resistance multiplied by the square of the current. If the current is sufficiently great, the heat will be generated at such a rate that the conductor rises in temperature so far that it becomes incandescent and radiates light. Attempts have been made to use platinum and platinum-iridium as the incandescent conductor, but these bodies are too expensive for general use, and besides, refractory though they are, they are not refractory enough to stand the high temperature required for economical incandescent lighting. Commercial success was not realized until very thin and very uniform threads or filaments of carbon were produced and enclosed in reservoirs of glass, from which the air was exhausted to the utmost possible limit. Such are the lamps made by Mr. Edison with which this building is lighted to-night. Let us examine the electrical properties of such a lamp. Here is a lamp intended to carry the same current as those overhead, but of half the resistance, selected because it leaves us a margin of electromotive force wherewith to vary our experiment. Into its circuit I am able to introduce a resistance for checking the current, composed of other incandescent lamps for convenience, but which I shall cover over that they may not distract your attention. As before, we have two galva-

nometers—one to measure the current passing through the
lamp, the other the difference of potential at its terminals.
First of all, we will introduce a considerable resistance;
you observe that, although the lamp gives some light, it
is feeble and red, indicating a low temperature. We take
our galvanometer readings. We now diminish the resistance. The lamp is now a little short of its standard intensity; with this current it would last 1,000 hours without
giving way. We again read the galvanometers. The resistance is diminished still further. You observe a great
increase of brightness, and the light is much whiter than
before. With this current the lamp would not last very
long. The results are given in Table IV.

TABLE IV.

Current Galvanometer.	Potential Galvanometer.	Ampères.	Volts.	Watts.	Resistance, Ohms.
5.2	12.8	0.38	37	14	97
6.0	14.3	0.44	41	18	93
11.5	23.4	0.84	68	57	81

There are three things I want you to notice in these
experiments: first, the light is whiter as the current increases; second, the quantity of light increases very much
faster than the power expended increases; and third, the
resistance of the carbon filament diminishes as its temperature increases, which is just the opposite of what we
should find with a metallic conductor. This resistance is
given in ohms in the last column. To the second point,
which has been very clearly put by Dr. Siemens in his British
Association address, I shall return in a minute or two.

The building is this evening lighted by about 200 lamps, each giving sixteen candles' light when 75 watts of power are developed in the lamp. To produce the same sixteen candles' light in ordinary flat flame gas burners would require between seven and eight cubic feet of gas per hour, contributing heat to the atmosphere at the rate of 3,400,000 foot-pounds per hour, equivalent to 1,250 watts; that is to say, equivalent gas lighting would heat the air nearly seventeen times as much as the incandescent lamps.

Look at it another way. Practically, about eight of these lamps take one indicated horse power in the engine to supply them. If the steam engine were replaced by a large gas engine this 1 h. p. would be supplied by 25 cubic feet of gas per hour, or by rather less; therefore by burning gas in a gas engine driving a dynamo, and using the electricity in the ordinary way in incandescent lamps, we can obtain more than five candles per cubic foot of gas, a result you would be puzzled to obtain in 16-candle gas burners. With arc lights instead of incandescent lamps many times as much light could be obtained.

At the present time, lighting by electricity in London must cost something more than lighting by gas. Let us see what are the prospects of reduction of this cost. Beginning with the engine and boiler, the electrician has no right to look forward to any marked and exceptional advance in their economy. Next comes the dynamo; the best of these are so good, converting 80 per cent. of the work done in driving the machine into electrical work outside the machine, that there is little room for economy in the conversion of mechanical into electrical energy; but the prime cost of the dynamo machine is sure to be greatly

reduced. Our hope of greatly increased economy must be
mainly based upon probable improvements in the incandescent lamp, and to this the greatest attention ought to
be directed. You have seen that a great economy of
power can be obtained by working the lamps at high pressure, but then they soon break down. In ordinary practice from 140 to 200 candles are obtained from a horse
power developed in the lamps, but for a short time I have
seen over 1,000 candles per horse power from incandescent
lamps. The problem, then, is so to improve the lamp in
detail that it will last a reasonable time when pressed to
that degree of efficiency. There is no theoretical bar to
such improvements, and it must be remembered that incandescent lamps have only been articles of commerce for
about three years, and already much has been done. If
such an improvement were realized, it would mean that
you would get five times as much light for a sovereign as
you can now. As things now stand, so soon as those who
supply electricity have reasonable facilities for reaching
their customers, electric lighting will succeed commercially
where other considerations than cost have weight. We
are sure of some considerable improvements in the lamps,
and there is a probability that these improvements may
go so far as to reduce the cost to one-fifth of what it now
is. I leave you to judge whether or not it is probable, nay,
almost certain, that lighting by electricity is the lighting
of the future.

DYNAMO-ELECTRIC MACHINERY.

THEORETICAL CONSTRUCTION OF CHARACTERISTIC CURVE.

OMITTING the inductive effects of the current in the armature itself, all the properties of a dynamo machine are most conveniently deduced from a statement of the relation between the magnetic field and the magnetizing force required to produce that field, or, which comes to the same thing but more frequently used in practice, the relation between the electromotive force of the machine at a stated speed and the current around the magnets. This relation given, it is easy to deduce what the result will be in all employments of the machine, whether as a motor or to produce a current through resistance, through an electric arc, or in charging accumulators; also the result of varying the winding of the machine, whether in the armature or magnets. The proper independent variable to choose for discussing the effect of a dynamo machine is the current around the magnets; and the primary relation it is necessary to know concerning the machine is the relation of the electromotive force of the armature to the magnet current. This primary relation may be expressed by a curve (Fig. 4, p. 22 *et seq.*, and Fig. 5, p. 26), now called the characteristic of the machine, and all consequences deduced therefrom graphically; or it may be expressed by stating the E.M.F. as an empirical function of the magnetizing current. Many such empirical formulæ have been proposed; as an instance we may mention that known as Fröhlich's, according to whom, if c be the current in the magnets, E the resulting E.M.F., $E = \dfrac{a\,c}{1 + b\,c}$. For some machines this formula is said to express observed results fairly accurately, but in our experience it does not sufficiently approximate to a straight line in the part of the curve near the origin. The character of the error in Fröhlich's formula is apparent by reference to Figs. 24 and 25, which give a series of observations on a dynamo machine, and for comparison therewith a hyperbola F, drawn as favorably as possible to accord with the observations.* Such empirical formulæ possess no advantage over the graphical method aided by algebraic processes, and tend to mask much that is of importance.

One purpose of the present investigation is to give an approximately complete construction of the characteristic curve of a dynamo of given form from the ordinary laws of electromagnetism and the known properties of iron, and to compare the result of such construction with the actual characteristic of the machine. The laws of electromagnetism needed are simply (Thomson, papers on " Elec-

---

* Added Aug. 17.—That Fröhlich's formula cannot be a thoroughly satisfactory expression of the characteristic of a dynamo machine is evident from the consideration that E should simply change its sign with c, that is, be an odd function of c. There should be a point of inflection in the characteristic curve at the origin. Another empirical formula, $\dfrac{E}{a} = \tan^{-1}\dfrac{c}{b}$, is free from this objection, but still fails to fully represent the approximation of the curve to a straight line on either side of the origin, and it is equally uninstructive with any other purely empirical formula.

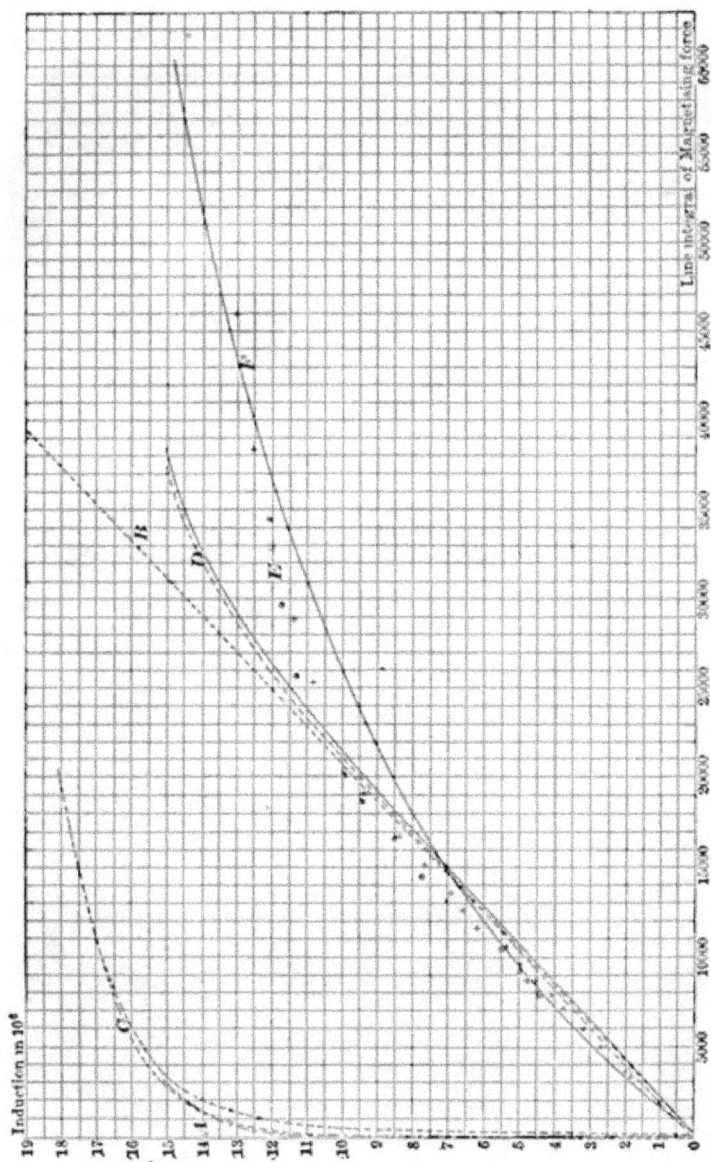

FIG. 24.—APPROXIMATE SYNTHESIS OF CHARACTERISTIC CURVE.

A, armature; B, air space; C, magnets; D, deduced curve; E, observed results, + ascending, ⊕ descending; F, Fröhlich's curve.

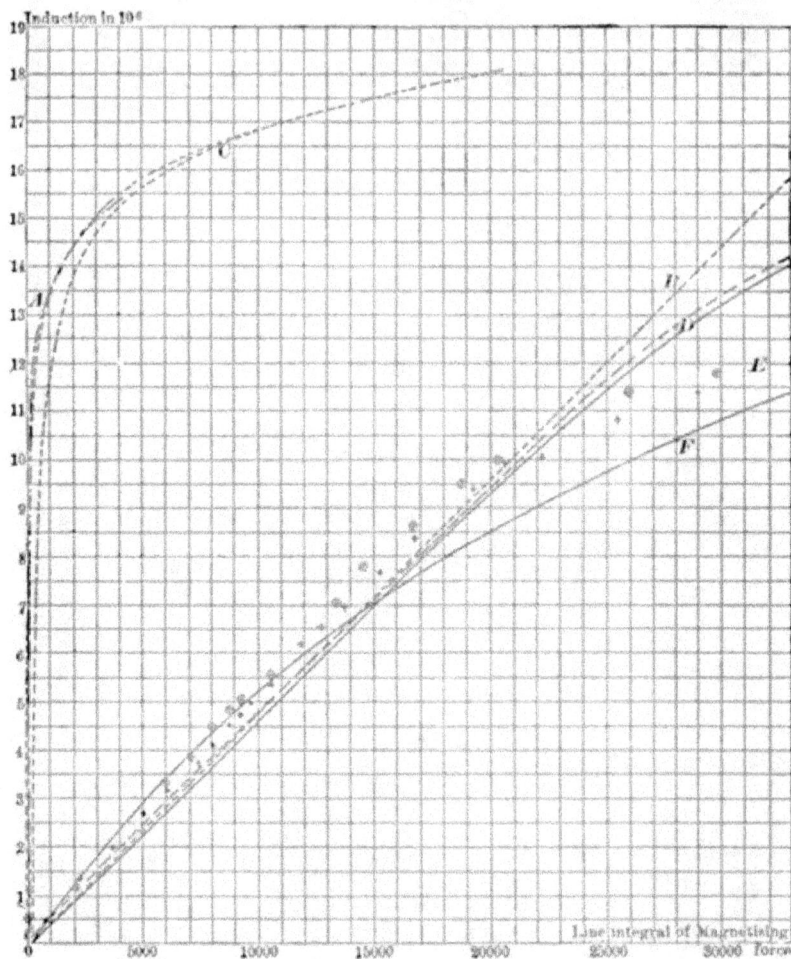

FIG. 25.—APPROXIMATE SYNTHESIS OF CHARACTERISTIC CURVE.
A, armature; *B*, air space; *C*, magnets; *D*, deduced curve; *E*, observed results. + ascending, ⊕ descending; *F*, Fröhlich's curve. This figure is the same as the left-hand part of Fig. 24, but on a larger scale.

trostatics and Magnetism;" Maxwell, "Electricity and Magnetism," vol. ii., pp. 24, 26, and 143), (1) that the line integral of magnetic force around any closed curve, whether in iron, in air, or in both, is equal to $4\pi n c$, where c is the current passing through the closed curve, and n is the number of times it passes through; (2) the solenoidal condition for magnetic induction, that is, if the lines of force or of induction be supposed drawn, then the induction through any tube of induction is the same for every section. Regarding the iron itself, we require to know from experiments on the material in any shape the relation between a, the induction, and α, the magnetic force at any point; for convenience write $a = f^{-1}(\alpha)$, or $\alpha = f(a)$. From these premises, without any further assumption, it is easy to see that a sufficiently powerful and laborious analysis would be capable of deducing the characteristic of any dynamo to any desired degree of accuracy. This we do not attempt, as, even if successful, the analysis would not be likely to throw any useful light on the practical problem. We shall calculate the characteristic, first making certain assumptions to simplify matters. We shall next point out the nature of the errors introduced by these assumptions, and make certain small corrections in the method to account for these sources of error, merely proving that the amount of these corrections is probable or deducing it from a separate experiment, and again compare the theoretical and the actual characteristic.

First Approximation.—Assume that by some miracle the tubes of magnetic induction are entirely confined to the iron excepting that they pass directly across from the bored faces of the pole pieces to the cylindrical face of the armature core. This, we shall find, introduces minor sources of error, affecting different parts of the characteristic curve to a material extent. Let I be total induction through the armature, A_1 the area of section of the iron of the armature, l_1 the mean length of lines of force in the armature; A_2 the area of each of the two spaces between core of armature and the pole pieces of the magnets, l_2 the distance between the core and the pole piece; A_3 the area of core of magnet, l_3 the total length of the magnets. All the tubes of induction which pass through the armature pass through the space A_2 and the magnet cores, and by our assumption there are no others. We now assume further that these tubes are uniformly distributed over these areas. The induction per square centimetre is then $\dfrac{I}{A_1}$ in the armature core, $\dfrac{I}{A_2}$ in the non-magnetic spaces, $\dfrac{I}{A_3}$ in the magnet cores; the corresponding magnetic forces per centimetre linear must be $f\left(\dfrac{I}{A_1}\right)$, $\dfrac{I}{A_2}$, $f\left(\dfrac{I}{A_3}\right)$. The line integral of magnetic force round a closed curve must be $l_1 f\left(\dfrac{I}{A_1}\right) + 2l_2 \dfrac{I}{A_2} + l_3 f\left(\dfrac{I}{A_3}\right)$. In this approximation we neglect the force required to magnetize pole pieces and other parts not within the magnet coils, to avoid complication. The equation of the characteristic curve is, then, $4\pi n c = l_1 f\left(\dfrac{I}{A_1}\right) + 2l_2 \dfrac{I}{A_2} + l_3 f\left(\dfrac{I}{A_3}\right)$. This curve is, of course, readily constructed graphically from the magnetic property of the material expressed by the curve $\alpha = f(a)$. In Figs. 24 and 25 curve A represents $x = l_1 f\left(\dfrac{y}{A_1}\right)$, the straight line B $x = 2l_1 \dfrac{y}{A_2}$, curve C $x = l_1 f\left(\dfrac{y}{A_2}\right)$, and curve D the calculated characteristic. When we compare this with an actual characteristic E, we shall see that, broadly speaking, it deviates from truth in

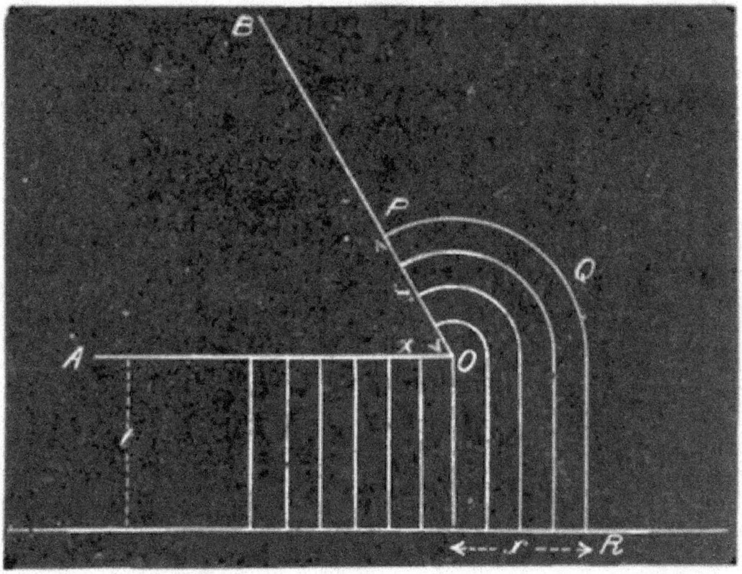

Fig. 26.

two respects: (1) it does not rise sufficiently rapidly at first; (2) it attains a higher maximum than is actually realized. Let us examine these errors in detail.

(1) The angle the characteristic makes with the axis of abscissæ near the origin is mainly determined by the line B (Fig. 26). We have in fact a very considerable exten-

sion of the area of the field beyond that which lies under
the bored face of the pole piece. The following consideration will show that the extension may be considerable:
Imagine an infinite plane slab, and parallel with it a second
slab cut off by a second plane making an angle α. We
want a rough idea of the extension of the area between the
plates by the spreading of the lines of induction beyond
the boundary. We know that the actual extension of the
area will be greater than we shall calculate it to be if we
prescribe an arbitrary distribution of lines of force other
than that which is consistent with Laplace's equation.

Assume, then, the lines of force to be segments of circles
centre O, and straight lines perpendicular to $O A$. The
induction along a line $P Q R$ will be $\dfrac{V}{(\pi - \alpha)x + t}$, V
being difference of potential between the planes; and the
added induction from $O P B$ will be

$$\int_0^x \frac{V\,d x}{(\pi - \alpha)x + t} = \frac{V}{\pi - \alpha} \log \frac{(\pi - \alpha)x + t}{t}.$$

Thus, if $\alpha = \dfrac{\pi}{2}$, we have for $x = t, 2t$, etc.,

x	$\dfrac{1}{\pi - \alpha} \log \dfrac{(\pi - \alpha)x + t}{t}$
t	0.599
$2t$	0.904
$3t$	1.109
$4t$	1.263
$5t$	1.387
$10t$	1.792

showing that the extension of the area of the field is likely to be considerable.

(2) The failure of the actual curve to reach the maximum indicated by approximate theory is because the theory assumes that all tubes of induction passing through the magnets pass also through the armature. Familiar observations round the pole pieces of the magnets show that this is not the case. If ν be the ratio of the total induction through the magnets to the induction in the armature, we must, in our expression for the line integral of magnetizing force, replace the term $f\left(\dfrac{I}{A_s}\right)$ by $f\left(\dfrac{\nu I}{A_s}\right)$: ν is not strictly a constant, as we shall see later; it is somewhat increased as I increases, owing to magnetization of the core of the armature, and it is also affected by the current in the armature. For our present purpose we treat it as constant.

There is yet another source of error which it is necessary to examine. Some part of the induction in the armature may pass through the shaft instead of through the iron plates. An idea of the amount of this disturbance may be readily obtained. Consider the closed curve $A\,B\,C\,D\,E\,F$: $A\,B$ and $F\,E\,D\,C$ are drawn along lines of force; $A\,F$ and $B\,C$ are orthogonal to lines of force (Fig. 27). Since this closed curve has no currents passing through it, the line integral of force around it is nil; therefore, neglecting force along $E\,D$, we have force along $A\,B$ equal to force along $F\,E$ and $D\,C$. In the machine presently described we may safely neglect the induction through the shaft; the error is comparable with the uncertainty as to the value of l_1; but in another machine, with magnets of much

greater section, the effect of the shaft would become very
sensible when the core is practically saturated.

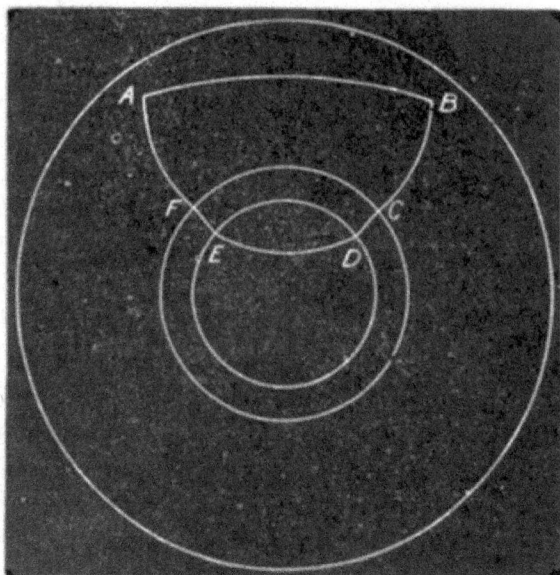

Fig. 27.

The amended formula now becomes

$$4\pi\,n\,c = l_{3}f\left(\frac{I}{A_{1}}\right) + 2l_{2}\frac{I}{A_{2}} + l_{3}f\left(\frac{\nu I}{A_{3}}\right) + l_{4}f\left(\frac{\nu I}{A_{4}}\right) + 2l_{5}f\left(\frac{I}{A_{5}}\right),$$

where l_{4} is the mean length of lines of force in the
wrought-iron yoke, A_{4} the area of the yoke, l_{5} and A_{5} corresponding quantities for the pole pieces, the last two
terms being introduced for the forces required to magnetize the yoke and the two pole pieces.

We now repeat the graphical method of construction exactly as before, the actual observations of induction in armature and current being plotted on the same diagram, Figs. 28 and 29, in which curve *G* represents the force required to magnetize the yoke, and curve *H* that required to magnetize the pole pieces. Before discussing these curves further, and comparing the results with those of actual observation, it may be convenient to describe the machine upon which the experiments have been made, confining the description strictly to so much as is pertinent to our present inquiry.

DESCRIPTION OF MACHINE.

The dynamo has a single magnetic circuit, consisting of two vertical limbs, extended at their lower extremities to form the pole pieces, and having their upper extremities connected by a yoke of rectangular section. Each limb, together with its pole piece, is formed of a single forging of wrought iron. These forgings, as also that for the yoke, are built up of hammered scrap and afterwards carefully annealed, and have a magnetic permeability but little inferior to the best Swedish charcoal iron. The yoke is held to the limbs by two bolts, the surfaces of contact being truly planed. In section the limb is oblong, with the corners rounded in order to facilitate the winding of the magnetizing coils. A zinc base, bolted to the bed-plate of the machine, supports the pole pieces.

The magnetizing coils are wound directly on the limbs, and consist of 11 layers on each limb, of copper wire 2.413 mms. diameter (No. 13, B.W.G.), making 3,260 convolu-

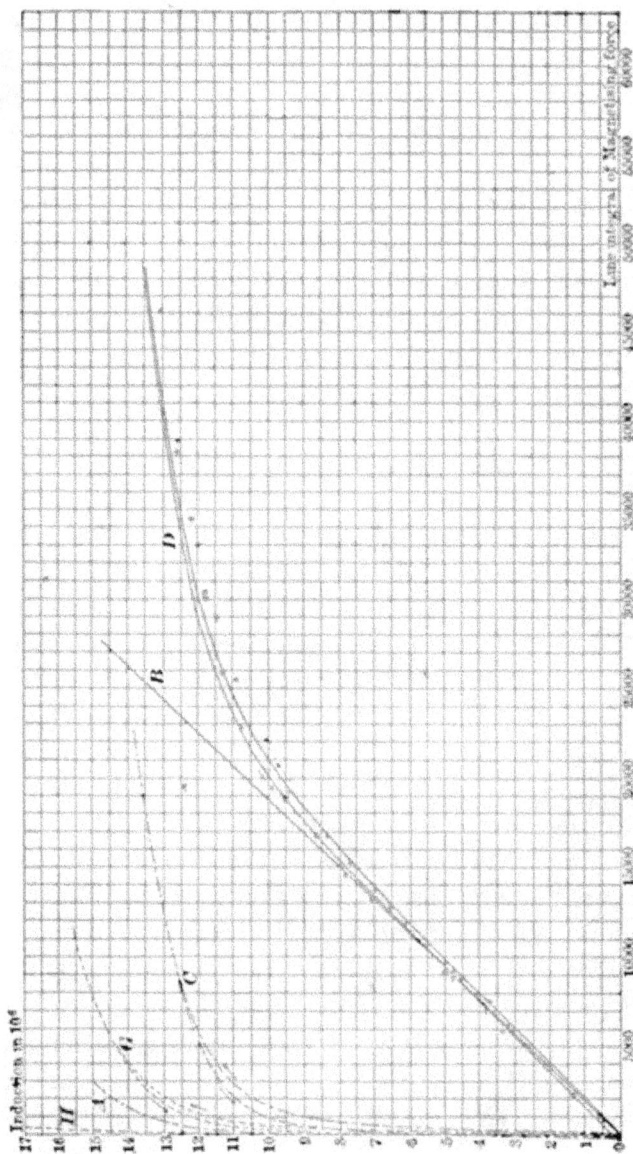

Fig. 28.—Correct Synthesis of Characteristic Curve.

A, armature; B, air space; C, magnets; D, calculated curve; E, observations, + ascending, ● descending; d, yoke; H, pole piece.

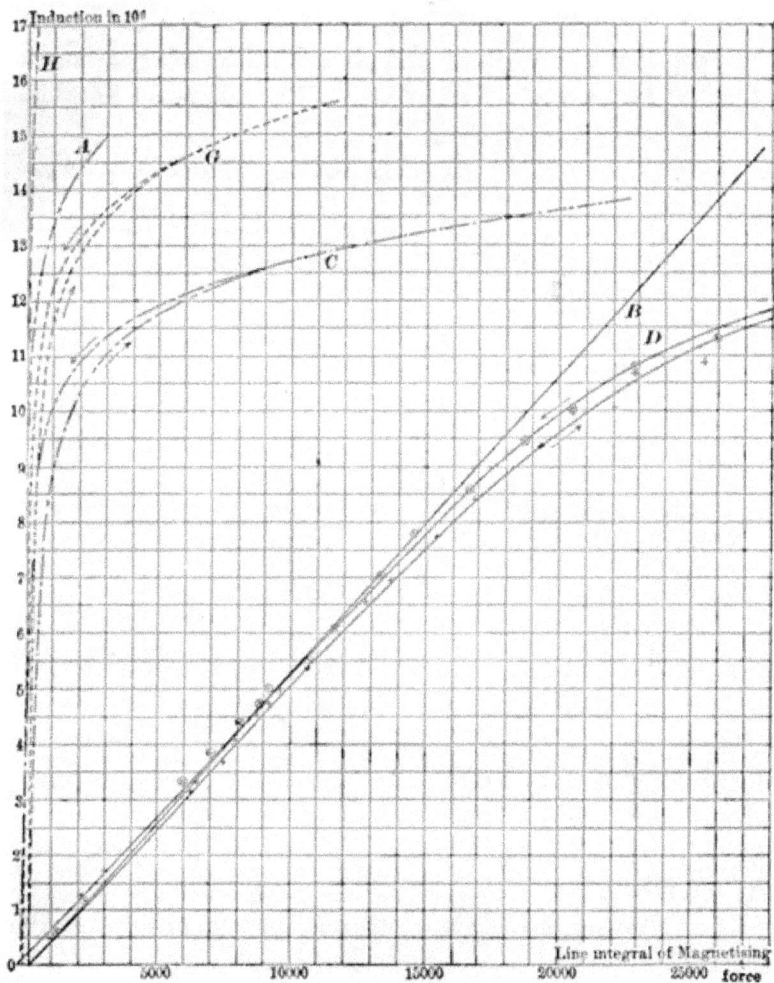

FIG. 29.—CORRECT SYNTHESIS OF CHARACTERISTIC CURVE.
A, armature; B, air space; C, magnets; D, calculated curve; E. observations, + ascending, ⊕ descending; G, yoke; H. pole piece. This figure is the same as the left-hand part of Fig. 28, but on a larger scale.

tions in all, the total length being approximately 4,570 metres. The pole pieces are bored to receive the armature, leaving a gap above and below, subtending an angle of 51° at the centre of the fields. The opposing surfaces of the gap are 8 mms. deep.

The following table gives the leading dimensions of the machine:

		cms.
Length of magnet limb....	=	45.7
Width of magnet limb	=	23.1
Breadth of magnet limb......................	=	44.45
Length of yoke	=	61 6
Width of yoke	=	48.3
Depth of yoke	-	23 2
Distances between centres of limbs	=	38.1
Bore of fields	=	27 5
Depth of pole piece	=	25 4
Width of pole piece measured parallel to the shaft	=	48.3
Thickness of zinc base	=	12 7
Width of gap	=	12 7

The armature is built up of about 1,000 iron plates, insulated one from another by sheets of paper, and held between two end plates, one of which is secured by a washer shrunk on to the shaft, and the other by a nut and locknut screwed on the shaft itself. The plates are cut from sheets of soft iron, having probably about the same magnetic permeability as the magnet cores. The shaft is of Bessemer steel, and is insulated before the plates are threaded on.

The following table gives the leading dimensions of the armature:

		cms.
Diameter of core.....................................	=	24.5
Diameter of internal hole........	=	7.62
Length of core over the end plates.....	=	50.8
Diameter of shaft.............. ....,	=	6.985

The core is wound longitudinally according to the
Hefner von Alteneck principle with 40 convolutions, each
consisting of 16 strands of wire 1.753 mm. diameter, the
convolutions being placed in two layers of 20 each.   The
commutator is formed of 40 copper bars, insulated with
mica, and the connections to the armature so made that
the plane of commutation in the commutator is horizontal
when no current is passing through the armature.

Fig. 30 shows a side elevation of the dynamo; Fig. 31 a

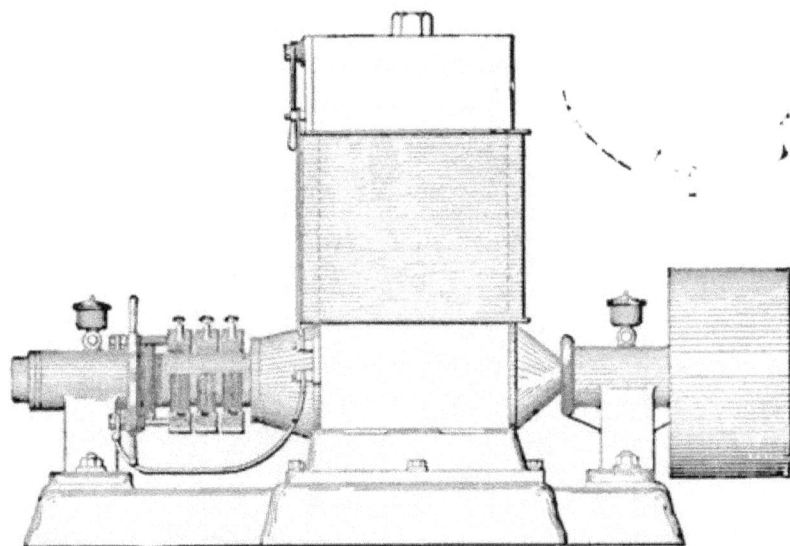

Fig 30.

cross section through the centres of the magnets; Fig 32
a section of the core of the armature, in a plane through
the axis of the shaft.

The dynamo is intended for a normal output of 105
volts 320 ampères at a speed of 750 revolutions per minute.

The resistance of the armature measured between opposite bars of the commutator is 0.009947 ohm, and of the mag-

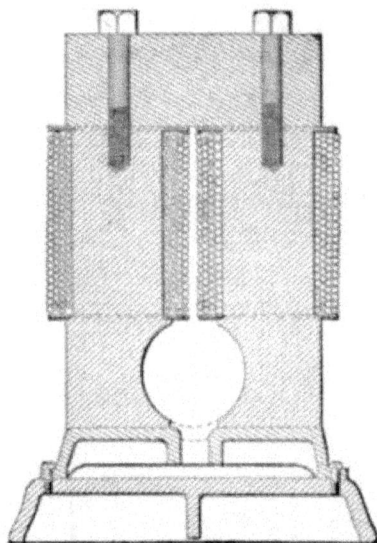

Fig. 31.

net coils 16.93 ohms, both at a temperature of 13.5° Centigrade; Lord Rayleigh's determination of the ohm being assumed.

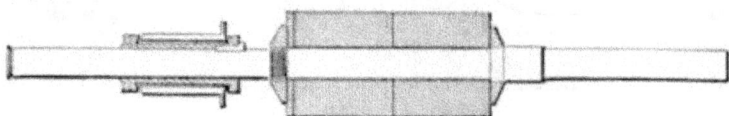

Fig. 32.

We have now to estimate the lengths and areas required in the synthesis of the characteristic curve.

A,;—from the length of the core of the armature (50.8 cms.) must be deducted 3.4 cms. for the thickness of insulating material between the plates; the resultant area is, on the other hand, as has already been stated, slightly augmented by the presence of the steel shaft. A_1 is taken as 810 sq. cms.

l_1;—this is assumed to be 13 cms., *i.e.*, slightly in excess of the shortest distance (12.6 cms.) between the pole pieces.

A_2;—the angle subtended by the bored face of the pole piece at the axis is 129°, the breadth of the pole piece is 48.3 cms., the diameter of the bore of the field is 27.5 cms., and, as already stated, the diameter of core 24.5 cms.; thus the area of pole piece is 1,513 sq. cms., and the area of 129° of the cylinder at the mean radius of 13.0 cms. is 1,410 sq. cms.; this value is taken for A_2 in the curves drawn in Figs. 24 and 25. In Figs. 28 and 29 A_2 is taken as 1,600, an allowance of 190 sq. cms. being made for the spreading of the field at the edges of the pole pieces, or $\dfrac{190}{10} = 1.2$ cm. all round the periphery, that is, $\dfrac{1.2}{1.5} = 0.8$ of the distance from iron of pole pieces to iron of core.

l_2 is 1.5 cm.

A_3 is a little uncertain, as the forgings are not tooled all over; it is here taken as 980 sq. cms., but this value may be slightly too high.

l_3 is 91.4 cms.

A_4 is 1,120 sq. cms.

l_4 is 49 cms., being measured along a quadrant from the centre of the magnet (see Fig. 33).

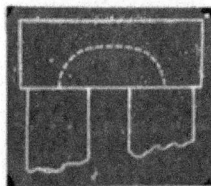

Fig. 33.

A_5 is 1,230 sq. cms., intermediate between the area of magnet and face of pole piece.

l_s is 11 cms.

ν was determined by experiment as described below, and its value is taken as 1.32; when the magnetizing current is more than 5.62 ampères its value should be a little greater.

The function $f(a)$ is taken from Hopkinson, *Phil. Trans.*, vol. clxxvi, 1885, p. 455; the wrought iron there referred to was not procured at the same time as, and its properties may differ to a certain extent from, the wrought iron of these magnets.

The curves now explain themselves: the abscissæ in each case represent the line integral of magnetizing force in the part of the magnetic circuit referred to; the ordinates, the number of lines of induction which also pass through the armature.

The results of the actual observations on the machine are indicated, those when the magnetizing force is increasing $+$, when it is decreasing $\oplus$. The measurements of the currents in the magnets which were separately excited, and of the potential difference between the brushes, the circuit being open, were made with Sir W. Thomson's graded galvanometers, standardized at the time of use. The irregularities of the observations are probably due to the variation of speed, the engine being not quite perfectly governed. The second construction exhibits quite as close an agreement between observation and calculation as could be expected; the deviation at high magnetizing forces is probably due to three causes—increase in the value of ν when the core of the armature is partially saturated, uncertainty as to the area A_s, difference in the quality of the iron. It is interesting to see how clearly theory predicts the difference between the ascending and descending curves of a dynamo. Consideration of the diagram proves that this machine is nearly perfect in its magnetic proportions. The core might be diminished without detriment by increasing the hole through it to a small, but very small, extent. Any reduction of area of magnets would be injurious; they might, indeed, be slightly increased with advantage. An increase in the length of the magnets would be very distinctly detrimental. Again, little advantage results from increasing the magnetizing force beyond the point at which the permeability of the iron of the magnets begins to rapidly diminish. For iron of the same quality as that of the machine under consideration, a magnetizing force of 2.6×10^{3} or 28.4 per centimetre is suitable. To get the same induction in other parts of the circuit, the diagram shows that for the air space a magnetizing force of 21×10^{3} is required, for the pole pieces 0.1×10^{3}, for the armature 0.2×10^{3}, for the yoke 0.6×10^{3}; making a total force required of 24.5×10^{3}. Any alteration in the length of the area of any portion of the magnetic circuit entails a corresponding alteration in the magnetizing forces required for that portion, at once deducible from the diagram. Similar machines must have the magnetizing forces proportional to the linear dimensions, and consequently, if the electromotive force of the machines is the same, the diameter of the wire of the magnet coils must be proportional to the linear dimensions. If the lengths of the several portions of the magnetic circuit remain the same, but the areas are similarly altered, the section of the wire must be altered in proportion to the alteration in the periphery of the section.

Around the middle of one of the magnet limbs a single coil of wire was taken, forming one complete convolution, and its ends connected to a Thomson's mirror galvanometer rendered fairly ballistic. If the circuit of the field magnets, while the exciting current is passing, be suddenly short circuited, the elongation of the galvanometer is a measure of the total induction within the core of the limbs, neglecting the residual magnetization. If the short circuit be suddenly removed, so that the current again passes round the field magnets, the elongation of the galvanometer will be equal in magnitude and opposite in direction.

The readings taken were:

Zero . . . . . . 71 left.
Deflection . . . . 332 " magnets made.
      "   . . . . 196 right; magnets short circuited.
      Hence, deflection to right = 267
              "   left = 261
              Mean deflection = 264

To determine the induction through the armature, the leads to the ballistic galvanometer were soldered to consecutive bars of the commutator, connected to that convolution of the armature which lay in the plane of commutation.

The readings taken were:

Zero 23   left.
Deflection 223   " magnets made.
    "   . . . . 176 ⎫ right; magnets short circuited.
    "   . . . . 178 ⎭

Hence, deflection to right and left = 200

It thus appears that out of 264 lines of force passing through the cores of the magnet limbs at their centre, 200 go through the core of the armature, whence ν equals 1.32. The magnetizing current round the fields during these experiments was 5.6 ampères.

EXPERIMENTS ON WASTE FIELD NOT PASSING THROUGH ARMATURE.

As in the determination of ν, a single convolution was taken around the middle of one of the limbs, and connected to the ballistic galvanometer; the deflections, when a current of 5.6 ampères was suddenly passed through the fields or short circuited, were:

Zero . . . . . . 34 left.
Deflection . . . . 148 " magnets made.
    "   . . . . 82 right; magnets short circuited.

Hence, deflection to right = 116
        "   left = 114
        Mean deflection = 115

I. Four convolutions were then wound round the zinc plate and the cast-iron bed in a vertical plane, passing through the axis of the armature; and the deflections noted were:

Zero . . . . 15 left.
Deflection . . . . 61 " magnets short circuited.
    "   . . . . . 40 right; magnets made.
Zero . . . . 11 left.
Deflection . . . . 64 " magnets short circuited.
    "   . . . . . . 36 right; magnets made.

Hence, deflection to right = 55
                  "   left = 46
and   "   right = 47
                  "   left = 53

in the two observations respectively, giving a mean = 50.25; or, reducing to one convolution, = 12.6.

II. A square wooden frame, 38 cms. × 38 cms., on which were wound ten convolutions, was then inserted between the magnet limbs, with one side resting on the armature, and an adjacent side projecting 5 cms. beyond the coils on the limbs, or about 7.6 cms. beyond the cores of the limbs. The deflections were:

Zero . . . . 34 left.
Deflection . . . . . 98 " magnets made.
    "   . . . . . 22 right; magnets short circuited.
    "   . . . . . 21 "   "   "
    "   . . . . . 81 left; magnets made.

Hence, deflection to right = 56
                  "   left = 64
and   "   right = 55
                  "   left = 47

in the two observations respectively, giving a mean = 55;
or, reducing to one convolution, = 5.5.

III. The same frame was raised a height of 6.35 cms.
above the armature in a vertical plane. The deflections
were:

Zero . . . . . .   21 left.
Deflection . . . .   98 " magnets made.
Zero . . . .   35 left.
Deflection . . . .   8 right; magnets short circuited.

Hence, deflection to left = 50
"   right = 43
and mean deflection = 46.5
or, reducing to one convolution, =   4.6

IV. The same frame was again lowered on the armature
and pushed inwards so as to lie symmetrically within the
space between the limbs. The deflections were:

Zero . . . .   32 right.
Deflection . . . . 112 "   magnets made.
"   . . . .   48 left; magnets short circuited.

Giving a mean of 80; or, reducing to one convolution,
= 8.0.

Let G represent the leakage through a vertical area
bounded by the armature and a line 7.6 cms. above the
armature and of the same width as the pole pieces; let R
be the remainder of the leakage between the limbs; then
II. and III. give

$$\frac{2}{3}G + \frac{2}{3}R = 5.5,$$

$$\frac{2}{3}R = 4.6;$$

whence

$$G = 1.35,$$
$$R = 6.9.$$

Again, IV. gives

$$\frac{5}{6}(G + R) = 8.0;$$

therefore

$$G + R = 9.6,$$

which shows an agreement as near as might be expected considering the rough nature of the experiment and that the leakage is assumed uniform over the areas considered.

We take

$$G = 1.6,$$
$$R = 8.0.$$

Reducing these losses to percentages we have

$$G = \frac{1.6}{115} \quad . \quad . \quad . \quad . \quad . \quad . \quad . \quad . \quad . \quad . \quad = 1.4 \text{ per cent.}$$

$$R = \frac{8.0}{115} \quad . \quad . \quad . \quad . \quad . \quad . \quad . \quad . \quad . \quad = 7.0 \quad ''$$

And from I. the leakage through the zinc plate and iron base	= 10.3	''
Hence the two gaps account for . . .	2.8	''
The zinc plate and iron base account for	10.3	''
And the area between the limbs . . .	7.0	''
Making a total loss accounted for . .	20.1	''
Out of an observed loss of	24.24	''

The leakage through the shaft and from pole piece to yoke, and one pole piece to the other by exterior lines, will account for the remainder.

EFFECT OF THE CURRENT IN THE ARMATURE.

The currents in the fixed coils around the magnets are
not the only magnetizing forces applied in a dynamo
machine; the currents in the moving coils of the armature
have also their effect on the resultant field. There are in
general two independent variables in a dynamo machine—
the current around the magnets and the current in the
armature; and the relation of E.M.F. to currents is fully
represented by a surface. In well-constructed machines
the effect of the latter is reduced to a minimum, but it
can be by no means neglected. When a section of the
armature coils is commutated, it must inevitably be
momentarily short circuited; and if at the time of commutation the field in which the section is moving is other
than feeble, a considerable current will arise in that section, accompanied by waste of power and destructive
sparking. It may be well at once to give an idea of the
possible magnitude of such effects. In the machine already described the mean E.M.F. in a section of the armature at a certain speed may be taken as 6 volts, its resistance 0.000995 ohm. Setting aside, then, for the moment
questions of self induction, if a section were commutated
at a time when it was in a field of one tenth part of the
mean intensity of the whole field, there would arise in that
section, while short circuited by the collecting brush, a
current of 600 ampères, four times the current when the
section is doing its normal work. The ideal adjustment
of the collecting brushes is such that during the time they
short circuit the sections of the armature the magnetic

forces shall just suffice to stop the current in the section, and to reverse it to the same current in the opposite direction.

Suppose the commutation occurs at an angle λ in advance of the symmetrical position between the fields, and

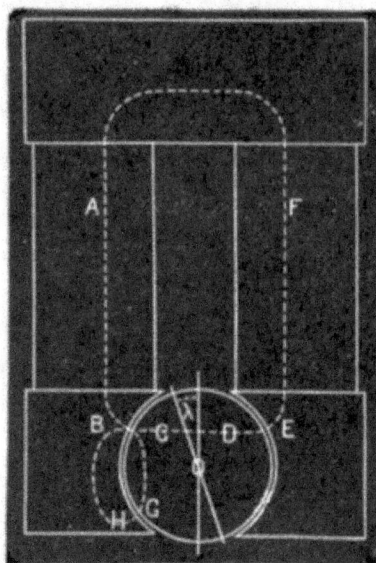

Fig. 34.

that the total current through the armature be C, reckoned positive in the direction of the resultant E.M.F. of the machine, *i.e.*, positive when the machine is used as a generator of electricity. Taking any closed line through magnets and armature, symmetrically drawn as $A\,B\,C\,D\,E\,F\,A$ (Fig. 34), it is obvious that the line integral of magnetic force is diminished by the current in the armature included between angle λ in front and angle

λ behind the plane of symmetry. If m be the number of convolutions of the armature, the value of this magnetizing force is $4\pi\, C\dfrac{m}{2}\dfrac{2\lambda}{\pi} = 4\lambda\, m\, C$ opposed to the magnetizing force of the fixed coils on the magnets. Thus if we know the lead of the brushes and the current in the armature we are at once in a position to calculate the effect on the electromotive force of the machine. A further effect of the current in the armature is a material disturbance in the distribution of the induction over the bored face of the pole piece; the force along $B\,C$ (Fig. 34) is by no means equal to that along $D\,E$. Draw the closed curve $B\,C\,G\,H\,B$: the line integral along $C\,G$ and $H\,B$ is negligible. Hence the difference between force $H\,G$ and $B\,C$ is equal to $4\pi\, C\dfrac{m}{2}\dfrac{\kappa}{\pi} = 2\kappa\, m\, C$, where κ is the angle $C\,O\,G$. This disturbance has no material effect upon the performance of the machine. But the current in the armature also distorts the arrangement of the comparatively weak field in the gap between the pole pieces, displacing the point of zero field in the direction of rotation in a generator and opposite to the direction of rotation in a motor; and it is due to this that the non-sparking point of the brushes is displaced. A satisfactory mathematical analysis of the displacement of the field in the gap between the pole pieces by the current in the armature would be more troublesome than an *à priori* analysis of the distribution of field in this space when the magnet current is the only magnetizing force. Owing to the fact that the armature is divided into a finite number of sections, there is a rapid diminution of the displacement of the field during the time that a section is being commutated, the diminution being recovered while the brush is in contact with only one bar of the commutator. The field thus oscillates slightly, owing to the disturbance caused by reversing the direction of the current in the successive sections of the armature. The number of oscillations in a Gramme armature or in a Siemens armature with an even number of sections will be ρm, where ρ is the number of revolutions per second; but in a Siemens armature with an odd number of sections it will be $2\rho m$.* This oscillation of the field is only another way of expressing the effect of the self induction of the section, but it must be remembered that if the self induction, multiplied by change of current, is expressed as a change in the field we must omit self induction as a separate term in our electrical equations. The precise lead to be given to the brushes in order to avoid sparking in any given case depends on many circumstances—the form and extent of the pole pieces, the number of sections in the armature, and the duration of the short circuit which the brushes cause in any section of the armature. The adjustment of the position of the collecting brushes is generally made by

* Added Aug. 17. Armatures with an odd number of convolutions are open to one theoretical objection, which would be a practical one if the number of convolutions were very small. The $2m+1$ convolutions constitute in themselves a closed circuit, having a resistance four times the mean actual resistance of the armature measured between the collecting brushes. When any one convolution is exactly in the middle of the field, the E.M.F. of the other $2m$ convolutions exactly balance, so that there is upon the closed circuit an E.M.F. due to the single convolution somewhat in excess of $\frac{1}{m}$ part of the actual E.M.F. of the machine. Thus there will be an alternating E.M.F. around the closed circuit of the armature capable of causing a considerable waste of power. This waste is materially checked by the self induction of the circuit.

hand at the discretion of the attendant, and is in some cases fixed once for all to suit an average condition of the machine. We shall, therefore, treat λ the lead as an independent variable, controlled by the attendant.

Let I be total induction through the armature, $I + I'$ total induction through the magnets, I' being the waste field. Let C be current in armature, c in the magnets. Let $g I'$ be the line integral of magnetic force from a point on one pole piece to a point on the other; the line being drawn external to the armature, g will be approximately constant. Omitting as comparatively unimportant the magnetizing force in the pole pieces and iron core of the armature, we have the following equations:—

$$4\lambda \, m \, C + 2l_{_2}\frac{I}{A_{_2}} - g \, I' = 0;$$

$$4\lambda \, m \, C + 2l_{_2}\frac{I}{A_{_2}} + l_{_1}f\left(\frac{I+I'}{A_{_1}}\right) = 4\pi \, n \, c.$$

When $C = O$, we observed

$$I = \frac{1}{\nu - 1} I';$$

whence

$$g = \frac{1}{\nu - 1}\frac{2l_{_2}}{A_{_2}};$$

eliminating I',

$$\frac{2l_{_2}}{\nu A_{_2}}\left(\nu I + 4\lambda \, m \, C\frac{A_{_2}}{2l_{_2}}(\nu - 1)\right) + l_{_1}f\left\{\frac{\nu I + 4\lambda m C.\frac{A_{_2}}{2l_{_2}}(\nu - 1)}{A_{_1}}\right\}$$

$$= 4\pi \, n \, c + 4\lambda \, m \, C\frac{\nu - 1}{\nu} - 4\lambda \, m \, C = 4\pi \, n \, c - 4\lambda \, m \, C\frac{1}{\nu}.$$

The characteristic curve when $C = O$ being $I = F(4 \pi n c)$, we may write the above as the equation of the characteristic surface thus:

$$I + \frac{\nu - 1}{\nu} 4 \lambda m C \frac{A_2}{2 l_2} = F\left(4 \pi n c - \frac{4 \lambda m C}{\nu}\right).$$

In applying this equation it must not be forgotten that the E. M. F. of the machine cannot be determined from I

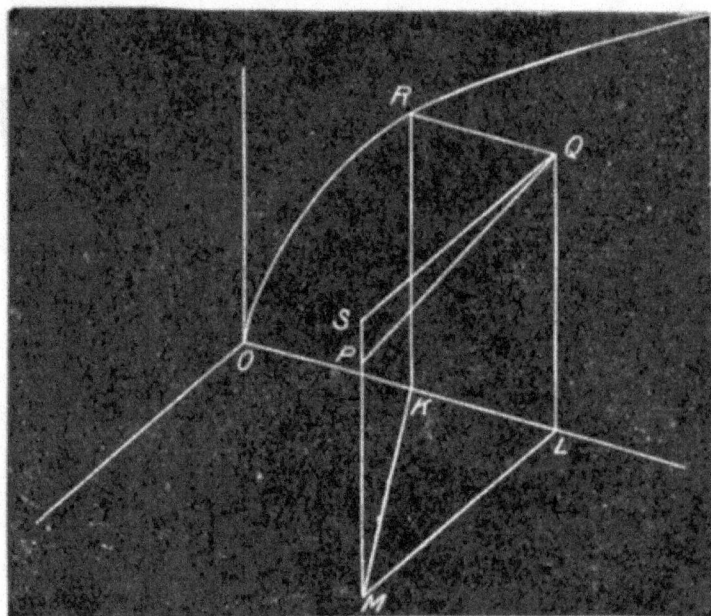

FIG. 35.

unless the commutation occurs at such a time that the coil being commutated embraces all, or nearly all, the lines of induction in the armature.

This equation enables the characteristic surface to be

constructed from the characteristic curve. Let $O\,L = 4\,\pi\,n\,c$ (Fig. 35), $L\,M = 4\,m\,\lambda\,C$; draw $M\,K$ so that $\dfrac{K\,L}{L\,M} = \dfrac{1}{\nu}$; through K draw ordinate $K\,R$, meeting characteristic curve in R; draw $R\,Q$ parallel to $O\,L$, meeting ordinate $Q\,L$ in Q; draw $Q\,S$ parallel to $L\,M$; draw $Q\,P$ so that $\dfrac{P\,S}{S\,Q}$ $= \dfrac{\nu - 1}{\nu} \cdot \dfrac{A_2}{2\,l_2}$. Then P is a point on the characteristic surface.

A very important problem is to deduce the characteristic curve of a series-wound machine from the normal characteristic; in this case $c = C$, and we have

$$I + \frac{\nu - 1}{\nu}\,4\,\lambda\,m\,C\frac{A_2}{2\,l_2} = F\left\{\left(4\,\pi\,n - \frac{4\,\lambda\,m}{\nu}\right)C\right\};$$

taking $P\,R$ (Fig. 36) as ordinate of any point in the normal characteristic, cut off $Q\,R$ equal to $\dfrac{\nu - 1}{\nu}\,4\,\lambda\,m\,C\dfrac{A_2}{2\,l_2}$ that is, draw $O\,Q$ so that

$$\tan Q\,O\,x = \frac{\nu - 1}{\nu}\,4\,\lambda\,m\,C\frac{A_2}{2\,l_2}\bigg/ 4\,\pi\left(n - \frac{m\,\lambda}{\nu\,\pi}\right)C$$

$$= \frac{\nu - 1}{\nu}\frac{A_2}{2\,l_2}\,\frac{\lambda\,m}{\pi\,n - \dfrac{\lambda\,m}{\nu}}.$$

Then $P\,Q$ will represent the induction corresponding to magnetizing force $4\,\pi\left(n - \dfrac{m\,\lambda}{\nu\,\pi}\right)C$. It is noteworthy that as the current C, and therefore OR, increases, PQ, the induction, will attain a maximum and afterwards diminish, vanish, and become negative. That in series-wound machines the E. M. F. has a maximum value has been many times observed. The cause lies in the existence of a waste field

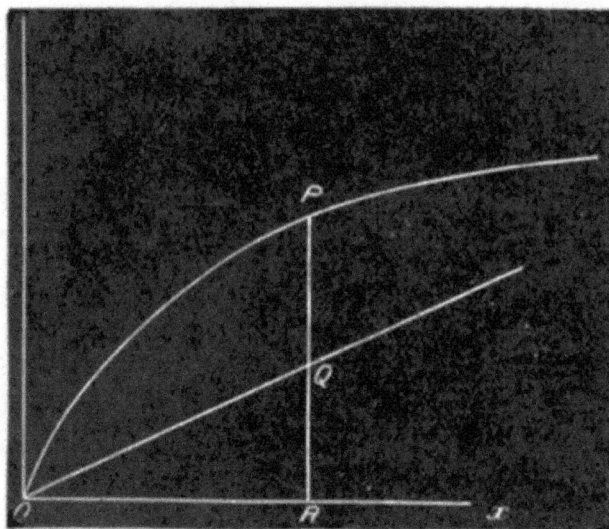

FIG. 36.

not passing through the armature, and in the saturation of the magnet core.

The effect of the current in the armature on the potential between the brushes of any machine is the same as that of an addition to the resistance of the armature proportional to the lead of the brushes and to the ratio of the waste field to the total field, combined with that of taking the main current $\dfrac{m}{\nu} \dfrac{\lambda}{\pi}$ times round the magnets in direction opposite to the current c. The preceding investigation tells the whole story of a dynamo machine, excepting only the relation of λ to C in order that the brushes may be so placed as to avoid sparking. The only constant or function which has to be determined experimentally for any particular machine is ν, the ratio of total to effective field; all the rest follows from the configuration of the iron and the known properties of the material.

The following illustrations of the effect of the current in the armature and the lead of the brushes are interesting. In both cases the magnet coils are supposed to be entirely disconnected, so that c is zero. First, let λ be negative, short circuit the brushes, and drive the machine at a certain speed; a large current will be produced, the current in the armature itself forming the magnet.* Second, let λ be positive, cause a current to pass through the armature: the armature will turn in the positive direction and will act as a motor capable of doing work. In either case, particularly the former, such use of the machine would not be practical, owing to violent sparking on the commutator. The following is a further illustration of the formula given above: If we could put up with the sparking which

---

* Added Aug. 17.—This experiment was tried upon a dynamo machine of construction generally similar to that shown in Figs. 30, 31, and 32, but with an armature of half the length intended in normal work to give 400 ampères, 50 volts, at 1,000 revolutions. The magnet coils were disconnected, and the terminals of the armature were connected through a Siemens electrodynamometer, and the machine was run at 1,380 revolutions. When the brushes were placed in the normal position ($\lambda = 0$) the current due to residual magnetism was 52 ampères. By giving the brushes a small positive lead the current was reduced to nearly zero. By giving the brushes a small negative lead a current of over 231 ampères, the maximum measured by the dynamometer, was obtained, and by varying the lead it was easy to maintain a steady current of any desired amount.

would ensue, it would be possible to make λ negative in a generator of electricity, and thereby obtain by the reactions of the armature itself all the results usually obtained by compound winding.

EFFICIENCY EXPERIMENTS.

Having discussed the relations subsisting between the configuration of the magnetic circuit of a dynamo machine and the induction obtained for given magnetizing forces, and having compared the results obtained by direct calculation with the results of actual observation on a particular machine, the construction of which we have described at length, it appeared of importance to determine the efficiency of the machine under consideration as a converter of energy, when used either as a generator of electricity or as a motor. An accurate determination of the mechanical power transmitted to a dynamo by a driving belt, or of the power given by a motor, presents formidable experimental difficulties. Moreover, if the mechanical power absorbed in driving the dynamo be measured directly, any error in measurement will involve an error of the same magnitude in the determination of the efficiency. To avoid this difficulty, we employed the following device:

Let two dynamos, approximately equal in dimensions and power, have their shafts coupled by a suitable coupling, which may serve also as a driving pulley; and let the electrical connections between the dynamos be made so that the one drives the other as a motor. If the combination be driven by a belt passing over the coupling pulley, the power transmitted by the belt is the waste in the two dyn-

amos and the connections between them. By suitably
varying the magnetic field of one of the dynamos, the
power passing between the two machines can be adjusted
as desired. If, then, the electrical power given out by the
generator is measured, and also the power transmitted by
the belt, the efficiency of the combination can be at once
determined. By this arrangement the measurement, which
presents experimental difficulties, viz., the power transmitted by the belt, is of a small quantity. Consequently
even a considerable error in the determination has but a
small effect on the ultimate result. On the other hand,
the measurement of the large quantity involved, viz., the
electrical power passing between the two machines, can
without difficulty be made with great accuracy.

The second machine was similar in all respects to that
already described, and each is intended for a normal output of 105 volts, 320 ampères, at a speed of 750 revolutions
per minute.

The power transmitted by the belt was measured by a
dynamometer of the Hefner-Alteneck type, the general
arrangement being as shown in the diagram, Fig. 37. A
is the driving pulley of the engine, B the driven coupling
of the dynamos; D, D are the guide pulleys of the dynamometer, carried on a double frame turning about the ful-
crum C, and supported by a spiral spring, the suspension
of which can be varied by a pair of differential pulley
blocks attached to a fixed support overhead. When a reading is made, the suspension of the spring is adjusted until
the index of the dynamometer comes to a fiducial mark on
a fixed scale; the extension of the spring is then read by a
second index attached to its upper extremity. F, F are two

fixed guide pulleys of the same diameter as the pulleys D, D, and having the same distance between their centres, in order that the two portions of the belt may be parallel and the sag as far as possible taken up. The normal from C

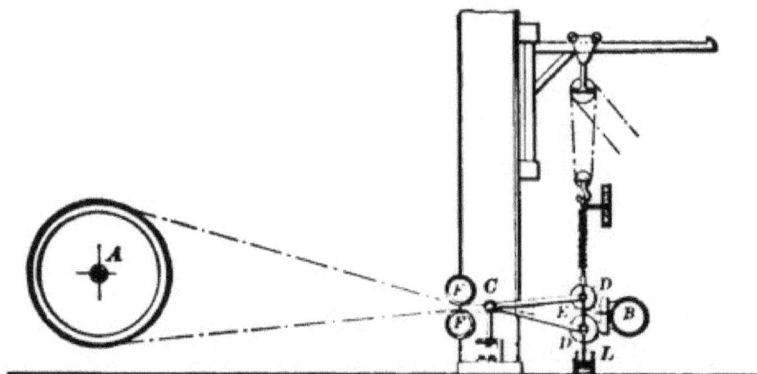

FIG. 37.

to the centre line of either portion of the belt between the pulley B and the guide pulleys $= 31.9$ cms. The normal from C to the centre line of either of the parallel portions of the belt $= 2.4$ cms., and from C to the centre line of the spring $= 92.7$ cms.

Take moments about C; then

$$\text{Tension of the belt} = \frac{92.7}{34.3} \times \text{tension of spring,}$$

$$= 2.7 \times \text{tension of spring.}$$

Also the diameter of the pulley $B = 33.6$ cms. and the thickness of the belt $= 1.6$ cm.

Hence the velocity of the centre of the belt in centimetres per second $= 1.845 \times$ revolutions of dynamo per minute, and, therefore,

Power transmitted by the belt in ergs per second $= 2.7 \times 1.845 \times 981 \times$ tension of spring $\times$ revolutions per minute, assuming the value of g to be 981.

We may more conveniently express the power in watts ($= 10^7$ ergs per second), and write

Power in watts $= 0.0004887 \times$ tension of spring
$$\times \text{ revolutions per minute.}$$

The potential between the terminals of the generator was measured by one of Sir William Thomson's graded galvanometers, previously standardized by a Clark's cell, which had been compared with other Clark's cells, of which the electromotive force was known by comparison with Lord Rayleigh's standard. The current between the two machines was measured by passing it through a known resistance, the difference of potential between the ends of the resistance being determined by direct comparison with the Clark's standard cell, according to Poggendorff's method. As experiments were made with currents of large magnitude, it was important that the temperature coefficient of the resistance should be as low as possible. To this end we found a resistance frame constructed of platinoid wire of great value. The temperature coefficient of this alloy is only 0.021 per cent. per degree Centigrade. (*Proc. Roy. Soc.*, vol. xxxviii, 1885, p. 265.)

The resistances of the armatures and magnets of the two machines are as follows:—

			Ohms.
Generator, . . .	armature,	. . .	0.009947
	magnets,	. . .	16.93
Motor, ...	armature,	. . .	0.009947
	magnets,	. . .	16.44

The resistance of the leads connecting the two machines was 0.00205 ohm, and of the standard resistance 0.00586 ohm.

In all determinations of resistance the value of the B. A.

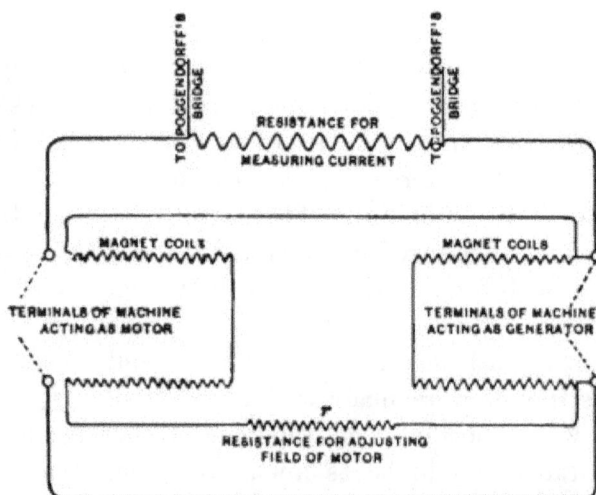

Fig. 38.

ohm was taken as 0.9867×10^9 C. G. S. units, according to Lord Rayleigh's determination.

The diagram, Fig. 38, shows the electrical connections between the two machines with the rheostat r inserted in the magnets of the motor dynamo.

In order to ascertain the friction of bending the belt round the pulley B, and of the journals of the dynamo, a preliminary experiment was made with the dynamometer. The combination was run at a speed of 814 revolutions per minute with the dynamos on open circuit, and the tension of the spring observed—9,979 grams. The engine was then reversed and the dynamos run at the same speed, and the tension of the spring again observed—3,629 grams. The difference of the two readings gives twice the power absorbed in friction, viz., 1,262 watts for the two machines, or 631 watts per machine. This is excluded entirely from the subsequent determinations of efficiency, as being a quantity dependent on such arbitrary conditions as the lubrication of the journals, the weight of the belt, and the angle it makes with the horizontal.

In Table V., column I. is the speed of the dynamos; column II. is the reading of the spring in grams; column III. is the power transmitted by the belt in watts; column IV. is the potential at the terminals of the generator; column V. is the current passing in the external circuit between the two machines; column VI. is the resistance introduced into the magnets of the motor by the rheostat; column VII. is the power absorbed in the armature of the generator; column VIII. is the power absorbed in the armature of the motor; column IX. is the power absorbed in the magnets of the generator; column X. is the power absorbed in the magnets of the motor; column XI. is the power absorbed in the connecting leads between the two dynamos, in the rheostat resistance r, and in the standard resistance used for measuring the current; column XII. is the total electrical power developed in the generator; col-

TABLE V.

	I. Revolutions per minute.	II. Grams.	III. Watts.	IV. Volts.	V. Ampères.	VI. Ohms.	VII. Watts.
1	810	8,392	3,322	129.1	21.6	1.39	18
2	801	9,299	3,640	127.2	72.0	1.39	75
3	811	11,113	4,405	125.8	150.0	2.72	267
4	808	10,433	4,119	124.4	186.0	2.72	807
5	792	10,660	4,124	116.5	211.0	2.72	499
6	798	16,897	6,589	110.6	351.0	4.59	1,309
7	764	17,690	6,605	110.12	358.0	4.09	1,360
8	766	17,804	6,665	110.6	360.0	4.59	1,375
9	778	16,556	6,394	102.3	369.0	4.09	1,436
10	756	20,412	7,541	96.8	446.0	4.59	2,070
11	808	9,526	3,765	119.3	36.8	2.72	22
12	802	3,855	1,512	113.5	No current	....	...
13	814	3,175	1,282	....	.. .	...	....

	VIII. Watts.	IX. Watts.	X. Watts.	XI. Watts.	XII. Watts.	XIII. Watts.	XIV. Watts.
1	4	984	861	77	4,720	691	5,411
2	51	955	837	112	11,096	805	11,901
3	223	935	709	295	20,896	988	21,884
4	344	914	693	348	25,256	691	25,949
5	443	801	608	453	26,590	660	27,250
6	1,222	722	455	1,101	41,433	890	42,323
7	1,268	716	473	1,131	42,087	828	42,915
8	1,289	722	455	1,152	42,194	836	43,380
9	1,354	618	408	1,178	40,314	650	40,964
10	1,979	554	348	1,670	46,244	459	46,704
11	13	841	637	116	5,998	1,066	7,064
12	....	....	....	....	......	756	......
13	....	....	...	....	......	631	......

umn XIII. is half the power absorbed by the combination less the known losses in the armatures, magnets, and external connections of the two machines; column XIV. is the total mechanical power given to the generator, being the sum of the powers given in columns XII. and XIII.

In Table VI. the percentage losses in the armature and magnets of the generator are given, as also the sum of all

TABLE VI.

	I. Per cent.	II. Per cent.	III. Per cent	IV. Per cent.	V. Per cent.	VI. Per cent.
1	0.24	18.20	12.76	68.8	57.28	39.40
2	0.63	8.93	6.76	84.58	82.99	70.19
3	1.22	4.27	4.52	90.00	90.15	81.13
4	1.53	3.52	2.66	92.28	92.65	85.49
5	1.83	2.94	2.415	92.80	93.12	86.42
6	3.09	1.71	2.10	93.10	93.30	86.86
7	3.17	1.67	1.93	93.23	93.39	87.07
8	3.17	1.67	1.93	93.23	93.43	87.10
9	3.51	1.75	1.59	93.39	93.50	87.32
10	4.43	1.19	0.98	93.39	93.36	87.19
11	0.35	21.9	15.1	72.65	65.77	47.78

other losses as obtained from column XIII. in Table V.; also the percentage efficiency of the generator, of the motor, and of the double conversion. Column I. is the percentage loss in the generator armature; column II. is the percentage loss in the generator magnets; column III. is the percentage sum of all other losses in the generator; column IV. is the percentage efficiency of the generator; column V. is the percentage efficiency of the motor; column VI. is the percentage efficiency of the double conversion.

In this series of experiments, in all cases from Nos. 1 to 10 inclusive, the brushes, both of the generator and motor, were set at the non-sparking point; but in No. 11 no lead was given to the brushes of the generator, and consequently there was violent sparking throughout the duration of the experiment.

In No. 12 the magnets were separately excited with a current giving 113.5 volts across their terminals. The power absorbed must be due entirely to local currents in the core of the armature and to the energy for the reversal of magnetization of the core twice in every revolution of the armature.

No. 13 gives the results of the experiments on the friction of the bearings and in bending the belt already referred to.

It will be observed that the figures in column XIII. are calculated by deducting the power absorbed in the armatures and magnets and by extraneous resistances from the total power given to the combination as measured by the dynamometer. They must therefore include all the energy dissipated in the core of the armature, whether in local currents or in the reversal of its magnetization; also the energy dissipated in local currents in the pole pieces, if such exist; also the energy spent in reversing the direction of the current in each convolution of the armature as they are successively short circuited by the brushes. Further, it will include the waste in all the connections of the machine from the commutator to its terminals and the friction of the brushes against the commutator. A separate experiment was made to determine the amount of this last constituent, but it was found to be too small to be capable of direct measurement by the dynamometer. Moreover, from the manner in which the figures in this column are deduced, any error in the dynamometric measurement will appear wholly in them. Since, undoubtedly, the first two components enumerated are the most important, and the conditions determining their amount are practically the same throughout the series, the close agreement of the figures in the column is a fair criterion of the accuracy of the observations. Probably 100 watts is the limit of error in any of the measurements. Such an error would affect the determination of the efficiency when the machines were working up to their full power by less than ¼ per cent.

It has been assumed that the sum of these losses is

equally divided between the two machines. This will not accurately represent the facts, as the intensities of the fields and the currents passing through the armatures differ to some extent in the two machines. The inequality, however, cannot amount to a great quantity, and if it diminishes the efficiency of the generator it will increase the efficiency of the motor by a like amount, and contrariwise. In No. 11 of the series the effect of the sparking at the brushes of the generator is very marked, the power wasted amounting to at least 250 watts.

If it be assumed that the dissipation of energy is the same whether the magnetization of the core is reversed by diminishing and increasing the intensity of magnetization without altering its direction, or whether it is reversed by turning round its direction without reducing its amount to zero, a direct approximation may be made to the value of this component. (J. Hopkinson, *Phil. Trans.*, vol. clxxvi, 1885, p. 455.)

The core has about 16,400 cubic centimetres of soft iron plates; hence loss in magnetizing and demagnetizing when the speed is 800 revolutions per minute $= 16,400 \times \frac{800}{60} \times 13,356$ ergs per second $= 292$ watts.

Referring to Table VI., it appears that the efficiency approaches a maximum when the current, passing externally between the two machines, is about 400 ampères. Let C be the current in the armature, ρ its resistance, W the power absorbed in all parts of the machine other than the armature; then, if the speed is constant, the efficiency is approximately $\dfrac{E C - W - C^2 \rho}{E C}$, where E is the electro-motive force. This is a maximum when $\dfrac{W}{C} + C \rho$ is a minimum, which occurs when $W = C^n\rho$; when the loss in the armature is equal to the sum of all other losses. For the machines under consideration the experimental results verify this deduction. But in actual practice the rate of generation of heat in the armature conductors, when a current of 400 ampères was passed for a long period, would be so great as to trench upon the margin of safety desirable in such machines. Of the total space, however, available for the disposition of the conductors, only about one fourth part is actually occupied by copper, the remainder being taken up with insulation and the interstices left by the round wire. If the space occupied by the copper should be increased to three fourths of the total space available, while the cooling surface remained the same, the current could be increased 75 per cent. and the efficiency increased 1.3 per cent. approximately, as all losses other than that in the armature wires would not be materially altered.

The loss in the magnets is also susceptible of reduction. It has already been shown that for a given configuration of the magnetic circuit and a given electromotive force the section of the wire of the magnet coils is determinate. The length is, however, arbitrary, since within limits the number of ampère convolutions is independent of the length. An increase in the length will cause a proportionate diminution in the power absorbed in the magnet coils. If the surface of the magnets is sufficient to dissipate all the heat generated, then the length of wire is properly determined by Sir William Thomson's rule that the cost of the energy absorbed must be equal to the continuing cost of the conductor.

APPENDIX.

(*Added Aug.* 17.)

Since the reading of the present communication experiments have been tried on machines having armatures wound on the plan of Gramme and with differently arranged magnets; the experiments were carried out in a closely similar manner to that already described.

DESCRIPTION OF MACHINES.

The construction of these machines is shown in Figs. 39,

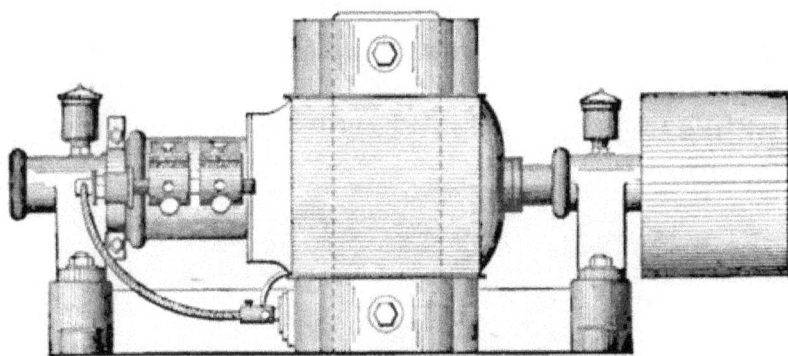

FIG. 39.

40, and 41, of which Fig. 39 shows an elevation, Fig. 40 a section through the magnets, Fig. 41 a longitudinal section of the armature. It will be observed that the magnetic circuit is divided. The pole pieces are of cast iron and are placed above and below the armature and are extended laterally. The magnet cores are of wrought iron of circular section and fit into the extensions of the cast iron

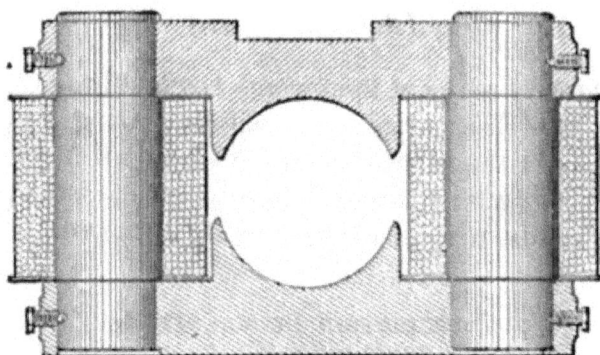

FIG. 40.

pole pieces, so that the area of contact of the cast iron is greater than the area of section of the magnet. The magnetizing coils consist of 2,196 convolutions on each limb

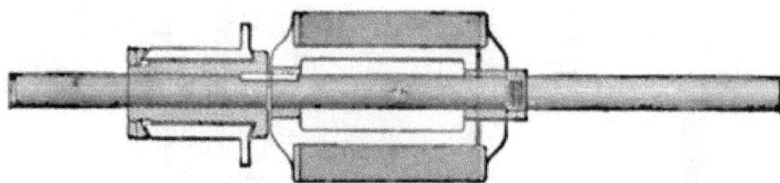

FIG. 41.

of copper wire, No. 17, B.W.G., in No. 1 machine, and 2,232 convolutions in No. 2 machine. The pole pieces are bored to receive the armature, leaving a gap on either side subtending an angle of 41° at the axis.

The bearings are carried upon an extension of the lower pole piece.

The following table gives the principal dimensions of the magnets in No. 1 machine:—

	cms.
Length of magnet limbs between pole pieces	26.0
Diameter of magnet limb	15.24
Bore of fields	25.7
Width of pole piece parallel to the shaft.	24.1
Width of gap between poles	8.6

The armature is built up of plates as in the machine already described, and is carried from the shaft by a brass frame between the arms of which the wires pass.

The principal dimensions are as follows:—

	cms.
Diameter of core	24.1
Diameter of hole through core	14.0
Length of core over end plates	24.1

The core is wound on Gramme's principle with 160 convolutions, each consisting of a single wire, No. 9, B.W.G., the wire lying on the outside of the armature in a single layer. The commutator has 40 bars.

This dynamo is compound wound, and is intended for a normal output of 105 volts, 130 ampères, at a speed of 1,050 revolutions per minute. The resistance of the armature is 0.047 ohm, and of the magnet shunt coils 26.87 ohms.

There is here no yoke, and consequently A_1 and l_1 do not appear in the equation.

It is necessary to bear in mind that the magnetizing force is that due to the convolutions on one limb, and that the areas are the sums of the areas of the two limbs. In calculating induction from E.M.F. it is also necessary to remember that two convolutions in a Gramme count as one in a Hefner-Alteneck armature.

A_1;—the section of the core is 245 sq. cms.; allowances for insulation reduce this to 220.5 sq. cms.

l_1;—this is assumed to be 10 cms., but it will be seen that an error in this value has a much more marked effect on the characteristic in this machine than in the other.

A_2;—the angle subtended by the bored face of the pole pieces is 139°; the mean of the radii of the pole pieces and the core is 12.45 cms. Hence the area of 139° of the cylinder of this radius is 768.3 sq. cms.; add to this a fringe of a width 0.8 of the distance from core to pole pieces, as already found necessary for the other machine, and we have 839.5 sq. cms. as the value of A_2.

l_2 is 0.8 cm.

A_3 is 365 sq. cms. (*i.e.*, the area of two magnet cores).

l_3 is 26.0 cms.

A_4 is taken to be 532 sq. cms., viz., double the smallest section of the pole piece.

l_4 is a very uncertain quantity; it is assumed to be 15 cms.

The expression already used requires slight modification. Inasmuch as the pole pieces are of cast iron, a different function must be used. Different constants for waste field must be used for the field, the pole pieces, and the magnet core. We write

$$4\pi n c = l_1 f\left(\frac{I}{A_1}\right) + 2 l_2 \frac{\nu_2 I}{A_2} + l_3 f\left(\frac{\nu_3 I}{A_3}\right) + 2 l_4 f\left(\frac{\nu_4 I}{A_4}\right).$$

The function f' is taken from Hopkinson, *Phil. Trans.*, vol. clxxvi, 1885, p. 455, Plate 52. ν_2, ν_3, and ν_4 were determined by experiment, as described below; their values are

$$\nu_2 = 1.05$$
$$\nu_3 = 1.18$$
$$\nu_4 = 1.49$$

Comparing the curves in Figs. 28 and 29 with that in Figs. 42 and 43, the most notable difference is that in the present case the armature core is more intensely magnetized than the magnet cores. No published experiments exist giving the magnetizing force required to produce the induction here observed in the armature core, amounting to a maximum of 20,000 per sq. cm. We might, however, make use of such experiments as the present to construct roughly the curve of magnetization of the material; thus we find that with this particular sample of iron a force of 740 per cm. is required to produce induction 20,000 per sq. cm.: this conclusion must be regarded as liable to considerable uncertainty.

The observations on the two machines are plotted together, but are distinguished from each other as indicated. They are, unfortunately, less accurate than those of Figs. 28 and 29, and are given here merely as illustrating the method of synthesis.

EXPERIMENTS TO DETERMINE ν_2, ν_3, AND ν_4.

The method was essentially the same as is described on pp. 96 to 99, and was only applied to No. 1 machine.

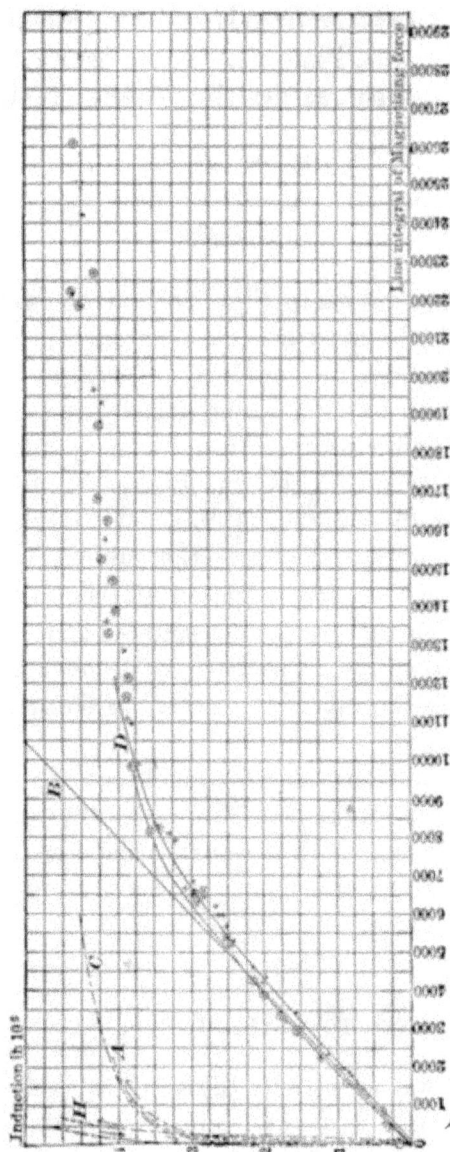

FIG. 42.—SYNTHESIS OF CHARACTERISTIC CURVE OF MACHINE WITH GRAMME ARMATURE.

Observed results: No. 1 machine, + ascending, ⊕ descending; No. 2 machine, × ascending, ● descending.

A, armature; B, air space; C, magnets; D, deduced curve; H, pole pieces

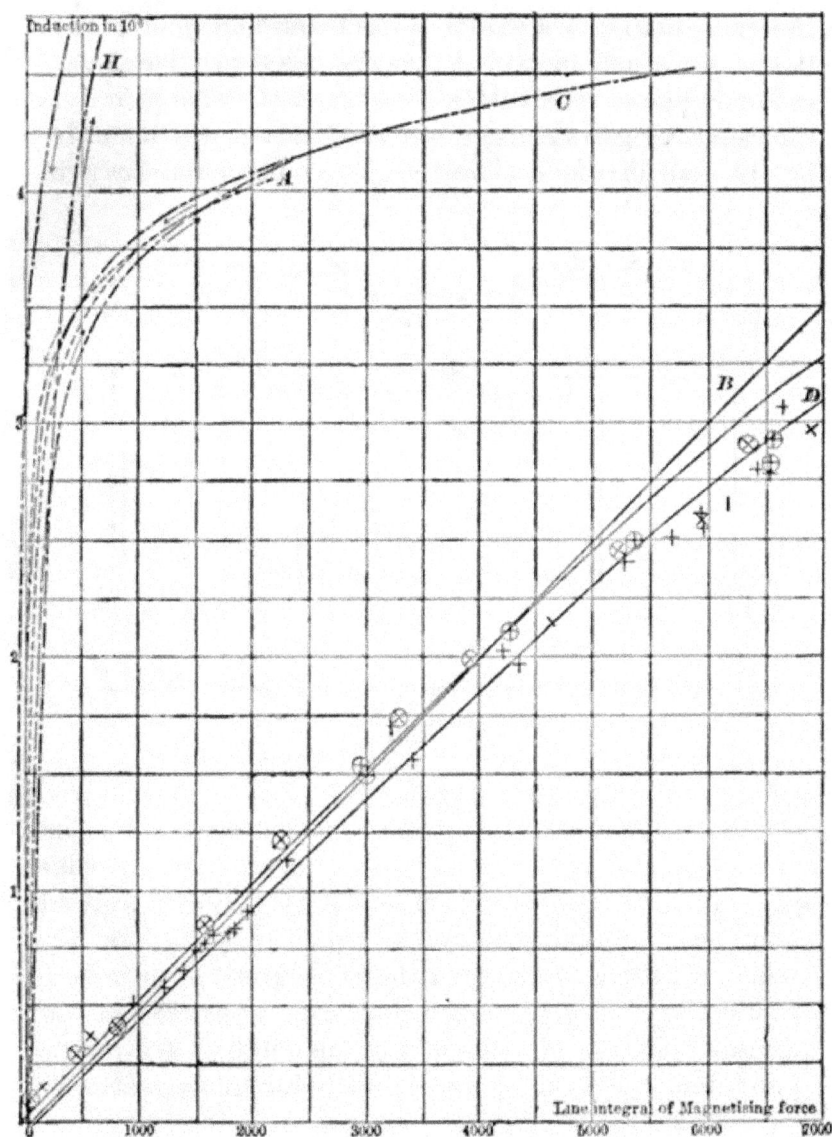

FIG. 43.—SYNTHESIS OF CHARACTERISTIC CURVE WITH GRAMME ARMATURE.
This figure is the same as the left-hand part of Fig. 42, but on a larger scale.

Referring to Fig. 44, a wire $A A$ was taken four times round
the middle of one limb of the magnet, a known current was
suddenly passed round the magnets, and the elongation of
the reflecting galvanometer was observed: it was found to
be 214 scale divisions, giving 107 as the induction through

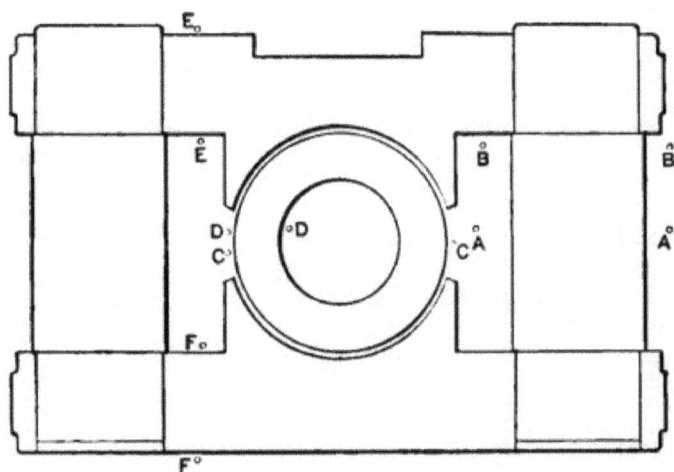

FIG. 44.

the two magnet limbs in terms of an arbitrary unit. The
coil was moved to the top of the limb as at $B B$; the elongation was reduced to 206, or 103 for the two limbs; we
take the mean induction in the magnet to be 105. A wire
was taken three times round the whole armature in a horizontal plane as at $C C$; the elongation observed was 222
divisions or 74 in terms of the same units. A wire was
taken four times round one half of the armature as at $D D$;

the elongation was 141, or induction in the iron of the armature 70.5, whence we have

$$\nu_3 = \frac{74}{70.5} = 1.05.$$

It may be well to recall here that ν_3 is essentially dependent on the intensity of the field; strictly the line B in Figs. 42 and 43 should not be straight, but slightly curved.

Four coils were taken round the upper pole piece at $E\,E$; the elongation was 159, giving 79.5 on the two sides. Coils at $F\,F$ give a higher result, 87.5, owing to the lines of induction which pass round by the bearings of the machine, and across to the upper ends of the magnets. ν_6 is taken to be $\dfrac{83.5}{70.5} = 1.18$.

EFFICIENCY EXPERIMENTS.

The method and instruments were those already described, pp. 110 to 112, excepting that the current was measured by a Thomson's graded galvanometer, which had been standardized against a Clark's cell in the position and at the time when used. The resistance of leading wires and galvanometer was 0.034 ohm, the series coils introduced for compounding the machines were also brought into use, and the losses due to their resistance (0.024 ohm) find a place in columns XII. and XIII. of Table VII., in which column I. is lead of brushes of the dynamo, positive for the generator, negative for the motor; column II., revolutions per minute; column III., deflection of spring in grams; column

TABLE VII.

	I. Degrees.	II. Revolutions.	III. Grams.	IV. Watts.	V. Volts.	VI. Ampères	VII. Ohms.	VIII. Watts.
1	17.5	1098	7711	4419	100.1	139.0	∞	955
2	5	1094	2722	1554	103.8	41.2	18.8	105
3*	0	1144	1814	1083	104.7	7.85	0	11

	IX. Watts.	X. Watts.	XI. Watts.	XII. Watts.	XIII. Watts.	XIV. Watts.	XV. Watts.	XVI. Watts.	XVII. Watts.
1	895	372	0	497	464	657	16,395	289	16,684
2	78	400	138	55	41	148	5,015	294	5,309
3*	3	408	408	6	1	2	1,637	128	1,765

* In this experiment the direction of the current had become reversed, and No. 2 machine was generator.

IV., watts by dynamometer; column V., volts at terminals of generator; column VI., ampères in external circuit; column VII., rheostat resistance; column VIII., watts in generator armature; column IX., watts in motor armature; column X., watts in generator shunt magnet coils; column XI., watts in motor shunt; column XII., watts in generator series magnet coils; column XIII., watts in motor series; column XIV., watts in external resistances; column XV., total electrical power of generator; column XVI., half the sum of losses unaccounted for; column XVII., total mechanical power applied to generator.

TABLE VIII.

	Generator Armature	Generator Shunt Coils.	Generator Series Coils.	Other Losses.	Efficiency of Generator.	Efficiency of Motor.	Efficiency of Double Conversion.
1	5.8	2.2	3.0	1.9	87.1	89.0	77.5
2	2.0	7.5	1.0	5.5	84.0	92.0	77.8

Table VIII. gives the losses and efficiencies as percentages in exactly the same way as in Table VI., excepting that another column is introduced for the loss in the series coils of the magnets of the generator.

The core of the armature contains about 6,500 cub. cms. of iron. Hence energy of magnetizing and demagnetizing when the speed = 1,100 revolutions per minute = 6,500 × $\frac{1,100}{60}$ × 13,356 in ergs per second = 159 watts.

DYNAMO-ELECTRIC MACHINERY.*

THE following is intended as the completion of a Paper† by Drs. J. and E. Hopkinson (*Phil. Trans.*, 1886).† The motive is to verify by experiment theoretical results concerning the effect of the currents in the armature of dynamo machines on the amount and distribution of the magnetic field which were given in that Paper, but which were left without verification. For the sake of completeness, part of the work is given over again.

The two dynamos experimented upon were constructed by Messrs. Siemens Brothers & Co., and are identical as far as it is possible to make them. They are mounted upon a common base plate, their axles being coupled together, and are referred to in this Paper respectively as No. 1 and No. 2.

Each dynamo has a single magnetic circuit consisting of two vertical limbs extended at their lower extremities to form the pole pieces, and having their upper extremities connected by a yoke of rectangular section. Each limb,

---

* It must not be supposed from his name not appearing in this short Paper that my brother, Dr. E. Hopkinson, had a minor part in the earlier Paper. He not only did the most laborious part of the experimental work, but contributed his proper share to whatever there may be of merit in the theoretical part of the Paper.—J. H.

† The Paper here referred to is that reprinted on pages 79 to 133 of this volume.

together with its pole piece, is formed of a single forging of wrought iron.   These forgings, as also that of the yoke, are built up of hammered scrap iron, and afterwards carefully annealed.   Gun-metal castings bolted to the base plate of the machine support the magnets.

The magnetizing coils on each limb consist of sixteen layers of copper wire 2 mms. in diameter, making a total of 3,968 convolutions for each machine.   The pole pieces are bored out to receive the armature, leaving a gap above and below subtending an angle of 68° at the centre of the shaft. The opposing surfaces of the gap are 1.4 cm. deep.

The following table gives the leading dimensions of the machine:—

	cms.
Length of magnet limb	66.04
Width of magnet limb	11.43
Breadth of magnet limb.	38.10
Length of yoke	38.10
Width of yoke	12.06
Depth of yoke	11.43
Distance between centres of limbs	23.50
Bore of fields	21.21
Depth of pole piece	20.32
Thickness of gun-metal base	10.80
Width of gap	12.06

The armature core is built up of soft iron disks, No. 24 B. W. G., which are held between two end plates screwed on the shaft.

The following table gives the leading dimensions of the armature:—

	cms.
Diameter of core	18.41
Diameter of shaft	4.76
Length of core	38.10

The core is wound longitudinally according to the Hefner von Alteneck principle with 208 bars made of copper strip, each 9 mms. deep by 1.8 mm. thick. The commutator is formed of fifty-two hard drawn copper segments insulated with mica, and the connections to the armature so made that the plane of commutation in the commutator is vertical when no current is passing through the armature.

Each dynamo is intended for a normal output of 80 ampères, 140 volts, at 880 revolutions per minute. The resistance of the armature measured between opposite bars of the commutator is 0.042 ohm, and of each magnet coil 13.3 ohms.

In the machine the armature core has a greater cross section than the magnet cores, and consequently the magnetizing force used therein may be neglected. The yoke has the same section as the magnet cores, and is therefore included therein, as is also the pole piece. The formula connecting the line integral of the magnetizing force and the induction takes the short form

$$4 \pi n c = 2 l_2 \frac{I}{A_2} + l_1 f \left(\frac{v I}{A_1} \right),^*$$

where

n is the number of turns round magnet;
c is the current round magnet in absolute measure;
l_2 the distance from iron of armature to rim of magnet;
A_2 the corrected area of field;
I the total induction through armature;
l_1 the mean length of lines of magnetic force in magnets;
A_1 the area of section of magnets;

---

* *Phil. Trans.*, 1886; page 88 of this volume.

ν the ratio of induction in magnets to induction in armature;

f the function which the magnetizing force is of the induction in the case of the machine actually taken from Dr. J. Hopkinson on the "Magnetization of Iron," *Phil. Trans.*, 1885, Figs. 4 and 5, Plate 47.

In estimating A_2 we take the mean of the diameter of the core and of the bore of the magnets 19.8 cms., and the angle subtended by the pole face 112°, and we add a fringe all round the area of the pole face equal in width to the distance of the core from the pole face. This is a wider fringe than was used in the earlier experiments,* because the form of the magnets differs slightly. The area so estimated is 906 sq. cms.

l_2 is taken to be 108.8 cms.

A_2 is 435.5 sq. cms.

ν was determined by the ballistic galvanometer to be 1.47. It is to be expected that, as the core is actually greater in area than the magnets, ν will be more nearly constant than in the earlier experiments. It was found to be constant within the limits of errors of observation.

Referring to Fig. 45, the curve C is the curve $x = l_2 f\left(\dfrac{\nu y}{A_2}\right)$, and the straight line B is the curve $x = 2\,l_2\,\dfrac{y}{A_2}$, while the full line D is the characteristic curve of the machine,

$$x = 2\,l_2\,\frac{y}{A_2} + l_2 f\left(\frac{\nu y}{A_2}\right),$$

as given by calculation.

---

* *Phil. Trans.*, 1886; page 95 of this volume.

The marks + indicate the results of actual observations on machine No. 1, **and the** marks 0 the results on machine **No. 2,** the total induction I being given **by the** equation:—

$$I = \frac{\text{potential difference in volts} \times 10^8}{208 \times \text{revolutions per second}}.$$

Experiments made upon the power taken to drive the machine under different conditions show that it takes about

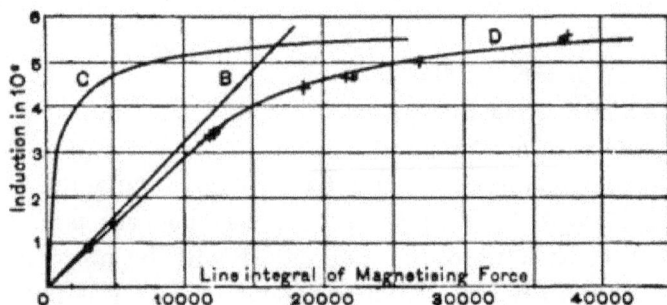

Fig. 45.

250 watts more power to turn the armature at 660 revolutions when the magnets are normally excited than when they are not excited at all. The volume of the core is 9,465 cub. cms., or in each complete cycle the loss per cubic centimetre is $\dfrac{250 \times 10^7}{11 \times 9,465} = 24,000$ ergs.

The loss by hysteresis is about 13,000 (*Phil. Trans.,* 1885, p. 463) if the reversals are made by variation of intensity of the magnetizing force and the iron is good wrought iron. This result is similar to that in the earlier

Paper,* where it is shown that the actual loss in the core, when magnetized, is greater than can be accounted for by the known value of hysteresis.

EFFECTS OF THE CURRENT IN THE ARMATURE.

Quoting from the Royal Society Paper [page 103 of this volume]. " The currents in the fixed coils around the magnets are not the only magnetizing forces applied in a dynamo machine—the currents in the moving coils of the armature have also their effect on the resultant field. There are in general two independent variables in a dynamo machine—the current around the magnets and the current in the armature; and the relation of E. M. F. to currents is fully represented by a surface. In well constructed machines the effect of the latter is reduced to a minimum, but it can be by no means neglected. When a section of the armature coils is commutated, it must inevitably be momentarily short circuited; and if at the time of commutation the field in which the section is moving is other than feeble, a considerable current will arise in that section, accompanied by waste of power and destructive sparking. . . .

" Suppose the commutation occurs at an angle λ in advance of the symmetrical position between the fields, and that the total current through the armature be C, reckoned positive in the direction of the resultant E. M. F. of the machine, i.e., positive when the machine is used as a generator of electricity. Taking any closed line through magnets and

---

* See page 121 of this volume.

armature, symmetrically drawn as $A\,B\,C\,D\,E\,F\,A$ [Fig. 46],
it is obvious that the line integral of magnetic force is diminished by the current in the armature included between
angle λ in front and angle λ behind the plane of symmetry. If m be the number of convolutions of the armature,
the value of this magnetizing force is $4\,\pi\,C\,\dfrac{m}{2}\dfrac{2\,\lambda}{\pi} = 4\,\lambda\,m\,C$
opposed to the magnetizing force of the fixed coils on the

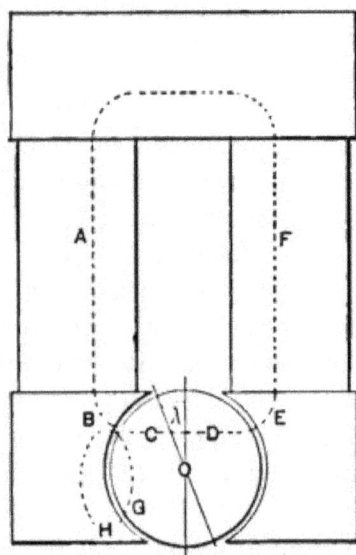

Fig. 46.

magnets. Thus if we know the lead of the brushes and the
current in the armature we are at once in a position to calculate the effect on the electromotive force of the machine.
A further effect of the current in the armature is a material
disturbance of the distribution of the induction over the

bored face of the pole piece; the force along BC [Fig. 46] is by no means equal to that along DE. Draw the closed curve $BCGHB$, the line integral along CG, and HB is negligible. Hence the difference between force HG and BC is equal to $4\pi C\dfrac{m}{2}\dfrac{\kappa}{\pi} = 2\kappa m C$, where κ is the angle COG."

To verify this formula is one of the principal objects of this Paper.

A pair of brushes having relatively fixed positions near together, and insulated from the frame and from one another, are carried upon a divided circle, and bear upon the commutator. The difference of potential between these brushes was measured in various positions round the commutator, the current in the armature, the potential difference of the main brushes, and the speed of the machine being also noted.

The results are given in Figs. 47, 48, 49, and 50, in which the ordinates are measured potential differences, and the abscissæ are angles turned through by the exploring brushes. The potential differences in Fig. 47 were measured by a Siemens voltmeter, and each ordinate is therefore somewhat smaller than the true value, owing to the time during which the exploring brushes were not actually in contact with the commutator segments. But this does not affect the results, because the area is reduced in the same proportion as the potential differences. In Figs. 48, 49, and 50 the potential differences were taken on one of Sir William Thomson's quadrant electrometers, and are correct.

Take Fig. 47, in which machine No. 1 is a generator. A

centimetre horizontally represents 10° of lead, and the ordinates represent differences of potential between the brushes. The area of the curve is 61.3 sq. cms., and represents 130 volts and a total field of $\frac{130}{104} \times \frac{1}{29} \times 10^{8}$ $= 4.31 \times 10^{8}$ lines of induction. This is, of course, not the actual field, which is 3 per cent. greater on account of the

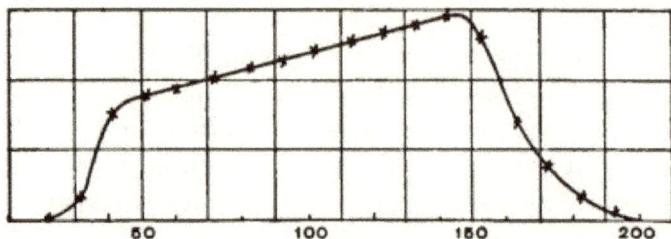

Fig. 47.

resistance of the armature, but is represented by an area 3 per cent. greater. An ordinate of 1 cm. will represent an induction of $\frac{4.31}{61.3} \times 10^{8} = 7.0 \times 10^{4}$ lines in 10°. The area of 10° is $39.5 \times 1.73 = 68.3$ sq. cms.* Hence an ordinate of 1 cm. represents an induction of 1,024 lines per square centimetre. The difference between ordinates at 50° and 140° is 2.5; hence the difference of induction is actually 2,560. Theoretically, we have $\kappa = \frac{1}{2} \pi m = 104 \, C = 9.4$. Therefore $2 \kappa m \, C = 3,072$, and this is the line integral of magnetizing force round the curve.

Let A be the induction at 50° and $A + \delta$ at 140°: these

---

* In calculating this area, the allowance for fringe at ends of armature is taken less than before, because the form of opposing faces differs.

also are the magnetizing forces. Hence $(A + \delta) 1.4 - A \, 1.4$
$= 2 \kappa m C$; $\delta = 2,200$, as against 2,560 actually observed.

Take Fig. 48, in which No. 2 machine is a motor. The
total field $= \dfrac{107}{104} \times \dfrac{1}{20} \times 10^{8} = 5.15 \times 10^{8}$ lines of induction. Since the area of the diagram is 53.5 sq. cms., an
ordinate of 1 cm. $= \dfrac{5.15}{53.5} \times 10^{8} = 96 \times 10^{4}$ lines of induc-

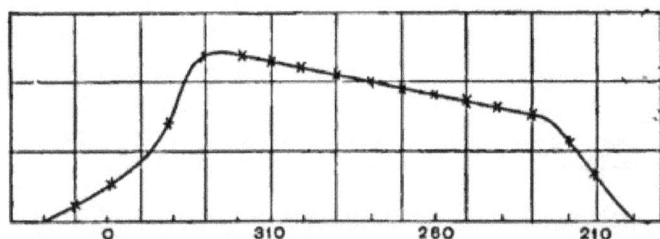

FIG. 48.

tion in 10°. Hence an ordinate of 1 cm. represents an
induction of $\dfrac{9.6 \times 10^{4}}{68.3} = 1,400$ lines per square centimetre.
The difference between ordinates at 320° and at 230° is 2.0;
hence the difference of induction is actually 2,800. Theo-
retically, we have $\dfrac{2 \kappa m C}{l} = \dfrac{3\frac{1}{4} \times 104 \times 11.4}{1.4} = 2,666$, as
against 2,800 actually observed.

In Fig. 49 No. 1 machine is a generator. The total field
$= \dfrac{52}{104} \times \dfrac{1}{12.6} \times 10^{8} = 3.97 \times 10^{8}$ lines. The area of the
diagram is 90.9 sq. cms., and therefore an ordinate of 1 cm.
$= \dfrac{3.97}{90.9} \times 10^{8} = 4.37 \times 10^{4}$ lines in 10°. Hence an ordi-

nate of 1 cm. represents an induction of $\dfrac{4.37 \times 10^4}{68.3} = 639$ lines per square centimetre. The difference between ordi-

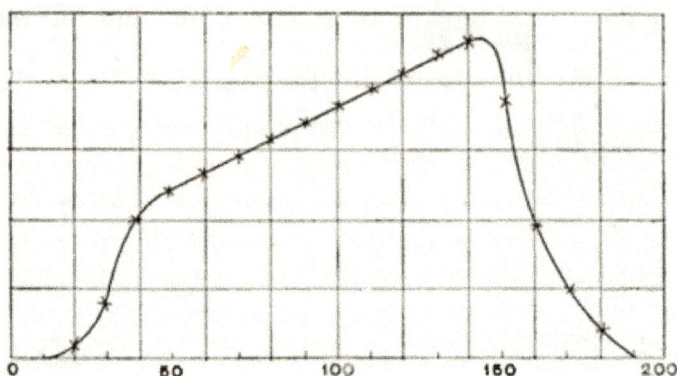

FIG. 49.

nates at 50° and at 140° is 4.5; hence the difference of induction is actually 2,877. Theoretically, we have $\dfrac{2 \kappa m C}{l}$

$$= \frac{3\frac{1}{4} \times 104 \times 12.9}{1.4} = 3,010, \text{ as against } 2,877.$$

In Fig. 50 No. 2 machine is a motor. The total field $= \dfrac{63.5}{104} \times \dfrac{1}{12.3} \times 10^7 = 4.96 \times 10^6$ lines. The area of the diagram is 112.2 sq. cms., and therefore an ordinate of 1 cm. $= \dfrac{4.96}{112.2} \times 10^6 = 4.42 \times 10^4$ lines in 10°. Hence an ordi-

nate of 1 cm. represents an induction of $\dfrac{4.42}{68.3} \times 10^4 = 647$ lines per square centimetre. The difference between ordinates at 323° and at 233° is 4.2; hence the difference

of induction is actually 2,718. Theoretically, we have
$$\frac{2\,\kappa\,m\,C}{l} = \frac{3\frac{1}{4} \times 104 \times 12.3}{1.4} = 2,870, \text{ as against 2,718 act-}$$
ually observed.

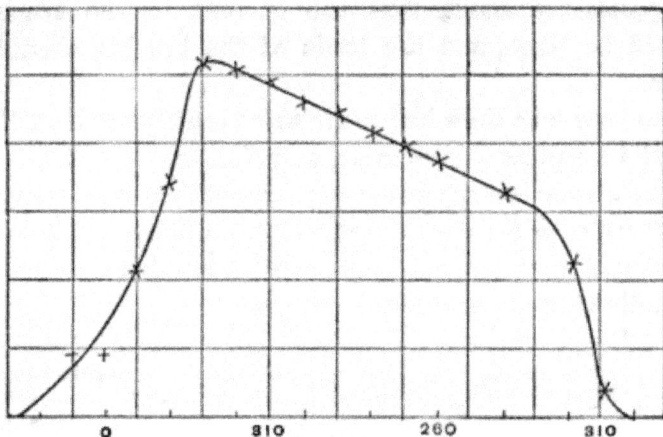

Fig. 50.

At page 108 of the preceding Paper on Dynamo-Electric Machinery it is shown that

$$I + \frac{\nu-1}{\nu}\,4\,\lambda\,m\,C\frac{A_2}{2\,l_2} = F\left(4\,\pi\,n\,c - \frac{4\,\lambda\,m\,C}{\nu}\right),$$

where $I = F(4\,\pi\,n\,c)$ is the characteristic curve when $C = 0$, and λ is the lead of the brushes.

The following is an endeavor to verify this formula. The potentials both upon the magnets and upon the brushes were taken by a Siemens voltmeter, and are rough. The speeds were taken by a Buss tachometer, and there is some uncertainty' about the precise lead of the brushes, owing to the difficulty in determining the precise position of the symmetrical position between the fields, and also to the width of the contacts on the commutator.

It was necessary, in order to obtain a marked effect of the armature reaction, that the magnet field should be comparatively small, that the current in the armature should be large, and the leads of the brushes should be large.

The two machines had their axles coupled so that No. 1 could be run as a generator, and No. 2 as a motor. The magnets were in each case coupled parallel, and excited by a battery each through an adjustable resistance. The two armatures were coupled in series with another battery, and the following observations were made:—

	Potential on Magnets in volts.	Potential on Brushes.	Speed per Minute.	Current in Ampères.	Lead of Brushes.
No. 1	24—24	66 - 67	880	102--103	26°
No. 2	29—29	86 - 84	880	102 - 103	29°

From which we infer:—

	Current in Magnets.	$4 \pi n c$.	Corrected Potential for Resistance of Armature.	Total Induction. I.
No. 1	1.78	8,900	70.8	2.80×10^6
No. 2	2.15	10,750	80.7	2.65×10^6

As there was uncertainty as to the precise accuracy of the measurements of potential, it appeared best to remeasure the potentials with no current through the armature with the Siemens voltmeter placed as in the last experiment. Each machine was therefore run on open circuit with its magnets excited, and its potential was measured.

	Potential on Magnets in volts.	Potential on Brushes.	Speed per Minute.	Potential at 880 Revs.
No. 1	25—25	90—90	880	90.0
No. 2	28—28	79—80	715—710	98.2

From which, since the formula is reduced to

$$I = \frac{A_2}{2\,l_2}(4\,\pi\,n\,c - 4\,\lambda\,m\,C),$$

the characteristic being practically straight, we infer:—

	Potential on Magnets.	Potential on Brushes.	Induction, $I = F(4\,\pi\,n\,c)$.
No. 1	24	86.4	2.82×10^6
No. 2	29	101.7	3.30×10^6

We have further:—

$\lambda = 0.45$ for No. 1; $\qquad$ $\lambda = 0.5$ for No. 2;

$\dfrac{4\,m\,C}{\nu} = 2{,}920;$ $\qquad$ $\dfrac{\nu-1}{\nu}\,4\,m\,C\,\dfrac{A_2}{2\,l_2} = 443{,}800.$

No.	$\dfrac{4\lambda mC}{\nu}$	$\dfrac{\nu-1}{\nu}4\lambda mC\dfrac{A_2}{2l_2}$	$4\pi nc - \dfrac{4\lambda mC}{\nu}$	$F\left(4\pi nc - \dfrac{4\lambda mC}{\nu}\right)$	$F\left(4\pi nc - \dfrac{4\lambda mC}{\nu}\right) - \dfrac{\nu-1}{\nu}4\lambda mC\dfrac{A_2}{2l_2}.$
1	1,314	199,700	7,586	2.41×10^6	2.21×10^6
2	1,460	221,900	9,290	2.90×10^6	2.68×10^6

It has already appeared that experiment gives for I in No. 1 2.3×10^6, and in No. 2 2.65×10^6. The difference is probably due to error in estimating the lead of the brushes, which is difficult, owing to uncertainty in the position of the neutral line on open circuit.

THEORY OF ALTERNATING CURRENTS, PARTICULARLY IN REFERENCE TO TWO ALTERNATE CURRENT MACHINES CONNECTED TO THE SAME CIRCUIT.

In my lecture on Electric Lighting, delivered before the Institution of Civil Engineers last year,* I considered the question of two alternate current dynamo machines connected to the same circuit, but having no rigid mechanical connection between them; and I showed that, if two such machines be coupled in series, they will tend to nullify each other's effect; if parallel, to add their effects.† The subject is one which already has practical importance and application, and may have much more in the future; it is also one suited for discussion, and upon which discussion is desirable. I therefore venture to bring before the Society what I said in my lecture—some other ways of looking at the same subject, and an experimental verification,

---

* This Paper is reprinted on pages 40 to 78 of this volume.

† *November 22, 1884.*—My attention has only to-day been called to a paper by Mr. Wilde, published by the Literary and Philosophical Society of Manchester, December 15, 1868, also *Philosophical Magazine*, January, 1869. Mr. Wilde fully describes observations of the synchronizing control between two or more alternate current machines connected together. I am sorry I did not know of his observations when I lectured before the Institution of Civil Engineers, that I might have given him the honor which was his due. If his paper had been known to those who have lately been working to produce large alternate current machines, it would have saved them both labor and money.

together with solutions of other problems requiring similar treatment.

The general explanation, amounting to proof so far as machines in series are concerned, is given in the following extract from my lecture:—

" There remains one point of great practical interest in connection with alternate current machines: How will they behave when two or more are coupled together to aid each other in doing the same work ? With galvanic batteries we know very well how to couple them, either in parallel circuit or in series, so that they shall aid, and not oppose, the effects of each other; but with alternate current machines, independently driven, it is not quite obvious what the result will be, for the polarity of each machine is constantly changing. Will two machines coupled together run independently of each other, or will one control the movement of the other in such wise that they settle down to conspire to produce the same effect, or will it be into mutual opposition ? It is obvious that a great deal turns upon the answer to this question, for in the general distribution of electric light it will be desirable to be able to supply the system of conductors from which the consumers draw by separate machines, which can be thrown in and out at pleasure. Now I know it is a common impression that alternate current machines cannot be worked together, and that it is almost a necessity to have one enormous machine to supply all the consumers drawing from one system of conductors. Let us see how the matter stands. Consider two machines independently driven, so as to have approximately the same periodic time and the same electromotive force. If these two machines

are to be worked together, they may be connected in one
of two ways: they may be in parallel circuit with regard
to the external conductor, as shown by the full line in
Fig. 51, that is, their currents may be added algebraically
and sent to the external circuit, or they may be coupled in
series, as shown by the dotted line, that is, the whole current may pass successively through the two machines, and
the electromotive force of the two machines may be added,

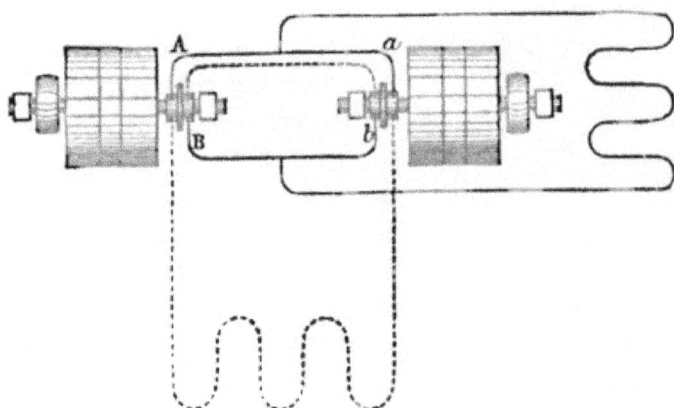

Fig. 51.

instead of their currents. The latter case is simpler. Let
us consider it first. I am going to show that if you couple
two such alternate current machines in series they will so
control each other's phase as to nullify each other, and
that you will get no effect from them; and, as a corollary
from that, I am going to show that if you couple them in
parallel circuit they will work perfectly well together, and
the currents they produce will be added; in fact, that you

cannot drive alternate current machines tandem, but that you may drive them as a pair, or, indeed, any number abreast. In diagram, Fig. 52, the horizontal line of abscissæ represents the time advancing from left to right; the full curves represent the electromotive forces of the two machines not supposed to be in the same phase. We want to see whether they will tend to get into the same phase or to get into opposite phases. Now, if the machines are coupled in series, the resultant electromotive force on the circuit will be the sum of the electromotive forces of the

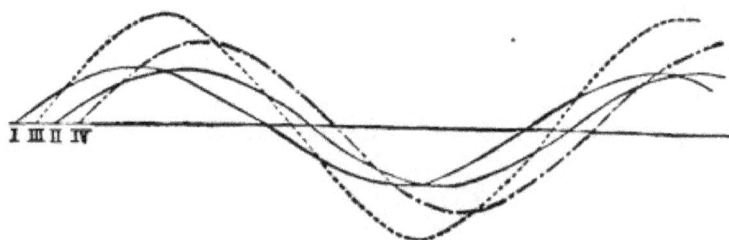

Fig. 52.

two machines. This resultant electromotive force is represented by the broken curve *III*; by what we have already seen in Formula IV. [p. 52, this volume], the phase of the current must lag behind the phase of the electromotive force, as is shown in the diagram by curve *IV*, thus ——.——.——. Now the work done in any machine is represented by the sum of the products of the currents and of the electromotive forces, and it is clear that, as the phase of the current is more near to the phase of the lagging machine *II* than to that of the leading machine *I*, the lagging machine must do more work in producing electricity than the leading machine; consequently its velocity will be retarded, and its retardation will go on until the two machines settle down into exactly opposite phases, when no current will pass. The moral, therefore, is, do not attempt to couple two independently driven alternate current machines in series. Now for the corollary: *A, B,* Fig. 51, represent the two terminals of an alternate current machine; *a, b,* the two terminals of another machine independently driven. *A* and *a* are connected together, and *B* and *b.* So regarded, the two machines are in series, and we have just proved that they will exactly oppose each other's effects, that is, when *A* is positive, *a* will be positive also; when *A* is negative, *a* is also negative. Now, connecting *A* and *a* through the comparatively high resistance of the external circuit with *B* and *b,* the current passing through that circuit will not much disturb, if at all, the relations of the two machines. Hence, when *A* is positive, *a* will be positive, and when *A* is negative, *a* will be negative also; precisely the condition required that the two machines may work together to send a current into the external circuit. You may, therefore, with confidence, attempt to run alternate current machines in parallel circuit for the purpose of producing any external effect. I might easily show that the same applies to a larger number; hence there is no more difficulty in feeding a system of conductors from a number of alternate current machines than there is in feeding it from a number of continuous current machines. A little care only is required that the machine shall be thrown in when it has attained something like its proper velocity. A further corollary is that alternate currents with alternate current machines as motors may theoretically be used for the transmission of power."*

Although the proof of this corollary regarding motors is similar to what we have just been going through, it may be instructive to give it. In the accompanying diagrams, Figs. 53 and 54, the full lines *I* and *II* represent the

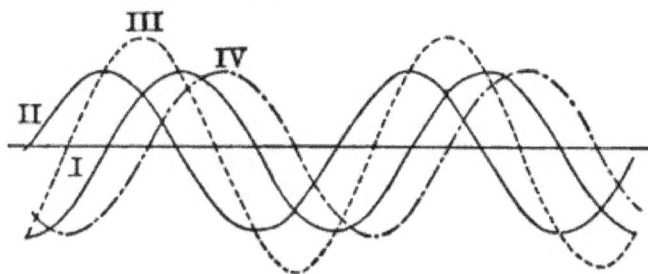

Fig. 53.

electromotive forces of the two machines (generator and receiver); the dotted line, curve *III* (. . . .), the resultant electromotive force; and the curve *IV*, the resulting current, each in terms of the time, as abscissæ. The only difference between the two diagrams is, that in Fig. 53 the two machines have equal electromotive forces, while in Fig. 54 *the receiving machine has double the electromotive force of the generator.* In both figures the receiving machine lags behind the phase of direct opposition to the generator by one quarter of a period, or something less. Now observe, the resultant electromotive force must be in

---

* "Of course in applying these conclusions it is necessary to remember that the machines only *tend* to control each other, and that the control of the motive power may be predominant and *compel* the two or more machines to run at different speeds."

phase behind the receiver, but in advance of the generator. Also observe, the current must be in phase behind the resultant electromotive force, and may be one quarter of a period behind, provided only the self induction be large enough compared with the resistance. The current will then be less than a quarter period behind the generator. This machine will do work upon the current, but the cur-

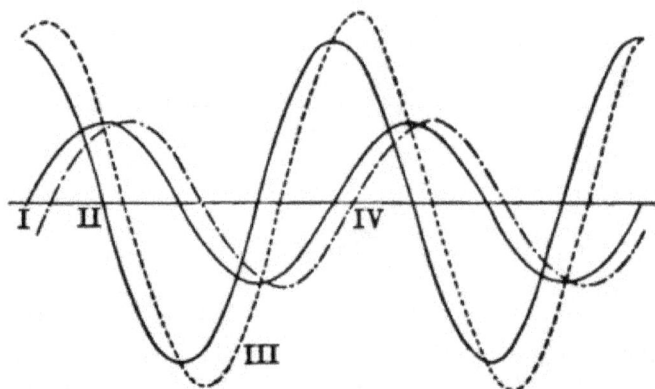

FIG. 54.

rent will be more than a quarter period behind the receiving machine; therefore in the receiver the current does work upon the machine.

The subject is illustrated by the following problems. Of course any of them may be treated more generally by considering the machines as unequal, or by introducing other periodic terms, but I do not see that this would throw more light on the subject:—

I. *Two alternate current machines, equal in all respects, are connected in series and independently driven at the same speed, to determine the current, etc., in each.*

Let γ be the coefficient of self induction of each, r the resistance, x the current at time t, and $E \sin \dfrac{2\pi}{T} (t + \tau)$ and $E \sin \dfrac{2\pi}{T} (t - \tau)$ the electromotive forces. Then regarding the coefficient of self induction as constant, which it is not exactly, and neglecting the effect of currents other than those in the copper wire, the equation of motion is

$$2\gamma x' + 2 r x = E \left\{ \sin \frac{2\pi}{T}(t + \tau) + \sin \frac{2\pi}{T}(t \overset{(-)}{\_} \tau) \right\},$$

or
$$\gamma x' + r x = E \sin \frac{2\pi t}{T} \cos \frac{2\pi \tau}{T};$$

whence

$$x = \frac{E \cos \dfrac{2\pi\tau}{T}}{r^2 + \left(\dfrac{2\pi\gamma}{T}\right)^2} \left\{ r \sin \frac{2\pi t}{T} + \frac{2\pi\gamma}{T} \cos \frac{2\pi t}{T} \right\}.$$

Work done by the leading machine per second

$$= \frac{E^2 \cos \dfrac{2\pi\tau}{T}}{2\left\{ r^2 + \left(\dfrac{2\pi\gamma}{T}\right)^2 \right\}} \cdot \left\{ r \cos \frac{2\pi\tau}{T} - \frac{2\pi\gamma}{T} \sin \frac{2\pi\tau}{T} \right\}$$

$$= \frac{E^2}{4\left\{ r^2 + \left(\dfrac{2\pi\gamma}{T}\right)^2 \right\}} \cdot \left\{ r\left(1 + \cos \frac{4\pi\tau}{T}\right) - \frac{2\pi\gamma}{T} \sin \frac{4\pi\tau}{T} \right\}.$$

From this it at once follows that the leading machine does least work, and will tend to increase its lead until $\tau = \dfrac{T}{4}$, when the two machines will neutralize each other, as already proved geometrically. *The leading machine may actually become a motor and do mechanical work, although its electromotive force is precisely equal to that of the following machine.*

Considering the important case when r is negligible, we have

$$x = -\frac{E \cos \frac{2\pi\tau}{T} \cdot \cos \frac{2\pi t}{T}}{\frac{2\pi\gamma}{T}},$$

rate of working $= \dfrac{E^2 \sin \frac{4\pi\tau}{T}}{4 \cdot \frac{2\pi\gamma}{T}}.$

This is the maximum when $\tau = \dfrac{T}{8}$, and then it is equal to the maximum work which can be obtained from either machine when connected to a resistance only, which occurs when that resistance is $\dfrac{2\pi\gamma}{T}$; the current, also, is the same as when the maximum work is being done on resistance, and is $\dfrac{1}{\sqrt 2}$ of the current the machine will give if short circuited. The difference of potential between the two leads connecting the machines, whether $r = 0$ or not, is $E \cos \dfrac{2\pi t}{T} \sin \dfrac{2\pi\tau}{T}$. If there be no work done on the receiving machine and $r = 0$, $\tau = \dfrac{T}{4}$, and the amplitude of the difference of potential between the leads is E; if, on the other hand, the maximum work is being transmitted, the potential measured will be $\dfrac{1}{\sqrt{2}}$ of that observed when either machine is run on open circuit.

II. *Two machines are coupled parallel and connected to an external circuit resistance R.*

Let x_1, x_2 be currents in the two machines. The external current will be $x_1 + x_2$, and consequently the difference of potential at the junction, $R (x_1 + x_2)$.

Let the electromotive forces of the two machines regarded in this case as connected parallel be $E \sin \dfrac{2 \pi (t \pm \tau)}{T}$, and let the self induction and resistance of each be 2γ and $2\,r$.

The equations of motion then are:

$$2 \gamma x'_1 + 2 r x_1 = E \sin \frac{2 \pi (t + \tau)}{T} - R (x_1 + x_2),$$

$$2 \gamma x'_2 + 2 r x_2 = E \sin \frac{2 \pi (t - \tau)}{T} - R (x_1 + x_2);$$

whence

$$\gamma (x'_1 + x'_2) + (R + r)(x_1 + x_2)$$

$$= E \sin \frac{2 \pi t}{T} \cdot \cos \frac{2 \pi \tau}{T},$$

and

$$\gamma (x'_1 - x'_2) + r (x_1 - x_2) = E \cos \frac{2 \pi t}{T} \sin \frac{2 \pi \tau}{T}.$$

Solving these,

$$x_1 + x_2 = \frac{E \cos \dfrac{2 \pi \tau}{T}}{(r + R)^2 + \left(\dfrac{2 \pi \gamma}{T}\right)^2}$$

$$\left\{ (r + R) \sin \frac{2 \pi t}{T} - \frac{2 \pi \gamma}{T} \cos \frac{2 \pi t}{T} \right\}$$

$$x_1 - x_2 = \frac{E \sin \dfrac{2 \pi \tau}{T}}{r^2 + \left(\dfrac{2 \pi \gamma}{T}\right)^2} \left\{ r \cos \frac{2 \pi t}{T} + \frac{2 \pi \gamma}{T} \sin \frac{2 \pi t}{T} \right\}.$$

Electrical work done by the leading machine

$$= \tfrac{1}{2} E \sin \frac{2 \pi (t + \tau)}{T} \{ x_1 + x_2 + (x_1 - x_2) \}$$

$$= \frac{1}{4} \frac{E^2}{(r + R)^2 + \left(\dfrac{2 \pi \gamma}{T}\right)^2} \left\{ (r + R) \cos^2 \frac{2 \pi \tau}{T} \right.$$

$$\left. - \frac{2 \pi \gamma}{T} \sin \frac{2 \pi \tau}{T} \cos \frac{2 \pi \tau}{T} \right\}$$

$$+ \frac{1}{4} \frac{E^2}{r^2 + \left(\dfrac{2 \pi \gamma}{T}\right)^2} \left\{ r \sin^2 \frac{2 \pi \tau}{T} \right.$$

$$\left. + \frac{2 \pi \gamma}{T} \sin \frac{2 \pi \tau}{T} \cos \frac{2 \pi \tau}{T} \right\}.$$

This expression shows that *the leading machine does most work in all cases.* Suppose r is small compared with R and $\dfrac{2\,\pi\,\gamma}{T}$, also that $R = \dfrac{2\,\pi\,\gamma}{T}$, we have the work done per second

$$= \frac{E^2}{8\,R}\left\{ \cos^2\frac{2\,\pi\,\tau}{T} + \sin\frac{2\,\pi\,\tau}{T}\cos\frac{2\,\pi\,\tau}{T} \right\}.$$

Make $\tau = -\dfrac{T}{8}$, and we see that the following machine will then do no work; when τ exceeds this, the following machine becomes a motor and absorbs electrical work.

III. *Suppose the terminals of an alternate current machine are connected to a pair of conductors, the difference of potential between which is completely controlled by connection with other alternate current machines.*

Let γ and R be the coefficient of self induction and the resistance of the machine and its own conductors up to the point at which the potential is completely controlled. Let the difference of potential of the main conductors be $A \sin \dfrac{2\,\pi\,t}{T}$, and let the electromotive force of the machine be $B \sin \dfrac{2\,\pi\,(t-\tau)}{T}$.

Equation of motion is

$$\gamma x' + R x = B \sin\frac{2\,\pi\,(t-\tau)}{T} - A \sin\frac{2\,\pi\,t}{T},$$

whence

$$
x = \frac{1}{R^2 + \left(\frac{2\pi\gamma}{T}\right)^2}\left[B\left\{R\sin\frac{2\pi(t-\tau)}{T} - \frac{2\pi\gamma}{T}\right.\right.
$$

$$
\left.\left.\cos\frac{2\pi(t-\tau)}{T}\right\} - A\left\{R\sin\frac{2\pi t}{T} - \frac{2\pi\gamma}{T}\cos\frac{2\pi t}{T}\right\}\right].
$$

Electrical work done by the machine in unit of time

$$
= x\,B\sin\frac{2\pi(t-\tau)}{T} = \frac{1}{R^2 + \left(\frac{2\pi\gamma}{T}\right)^2}
$$

$$
\left[\frac{B^2 R}{2} - \frac{A B}{2}\left\{R\cos\frac{2\pi\tau}{T} + \frac{2\pi\gamma}{T}\sin\frac{2\pi\tau}{T}\right\}\right].
$$

If τ be positive, that is, if machine be lagging in its phase, work done is less than if it be negative; hence τ will tend to zero, or the machine will tend to adjust itself to add its currents to that of the system of conductors. The machine may act as a motor *even though its electromotive force be greater than that of the system,* for let

$$
\frac{R}{\dfrac{2\pi\gamma}{T}} = \tan\frac{2\pi\phi}{T},
$$

work (electric) done by machine

$$= \frac{B^2 R}{2 \left\{ R^2 + \left(\frac{2\pi\gamma}{T}\right)^2 \right\}}$$

$$- \frac{A B}{2 \left\{ R^2 + \left(\frac{2\pi\gamma}{T}\right)^2 \right\}^{\frac{1}{2}}} \sin \frac{2\pi(\phi+\tau)}{T};$$

this has a minimum value when $\phi + \tau = \frac{T}{4}$, and then the mechanical work done by machine or electrical work received by the machine

$$= \frac{B}{2 \left\{ R^2 + \left(\frac{2\pi\gamma}{T}\right)^2 \right\}^{\frac{1}{2}}} \left\{ A - \frac{R B}{\left\{ R^2 + \left(\frac{2\pi\gamma}{T}\right)^2 \right\}^{\frac{1}{2}}} \right\},$$

and this is positive, provided

$$\frac{A}{B} > \frac{R}{\left\{ R^2 + \left(\frac{2\pi\gamma}{T}\right)^2 \right\}^{\frac{1}{2}}}.$$

There are two or three other problems of sufficient interest to make it worth while giving them here. although not directly relating to alternate current machines coupled together.

IV. *To determine the law of an alternate current through an electric arc.*

It has been shown by Joubert that in an arc the difference of potential is of approximately constant numerical

value, reversing its value discontinuously with the reversal
of the current, probably at the instant of reversal of current. We shall assume, then, that there is in the arc a
constant electromotive force, A, always opposed to the
current, except when the current ceases, and that then its
value is zero.

The equation of motion is

$$\gamma x' + R x = E \sin \frac{2\pi t}{T} \mp A,$$

the negative sign being taken when x is $+ \infty$, the positive
when x is negative. Solving generally,

$$x = \mp \frac{A}{R} + \frac{E}{\left(\frac{2\pi\gamma}{T}\right)^2 + R^2} \left(-\frac{2\pi\gamma}{T} \cos \frac{2\pi t}{T} \right.$$
$$\left. + R \sin \frac{2\pi t}{T} \right) + C e^{-\frac{R}{\gamma} t},$$

This equation will continuously hold good for a half period
from $x = 0$ to $x = 0$ again, but at each half period the
arbitrary constant C is changed with the sudden change of
sign of A. It is determined by the consideration that if,
for a certain value t_0 of t, x should vanish, it shall vanish
again when $t = t_0 + \frac{T}{2}$. This applies to the case when E
is sufficiently large, as is practically the case; but if the
current should cease for a finite time this condition will be
varied, and instead of it we have the condition $x = 0$ when
$E \sin \frac{2\pi t}{T} = A$. This latter case I do not propose to
consider further.

Let

$$\frac{2\pi\gamma}{RT} = \tan\frac{2\pi t_1}{T};$$

$$x = \mp\frac{A}{R}$$

$$+ \frac{E}{\sqrt{\left\{R^2 + \left(\frac{2\pi\gamma}{T}\right)^2\right\}}} \sin\frac{2\pi(t - t_1)}{T} + Ce^{-\frac{R}{\gamma}t}.$$

Putting $\iota = t_0$ and $t = t_0 + \dfrac{T}{2}$, we have

$$0 = -\frac{A}{R}$$

$$+ \frac{E}{\sqrt{\left\{R^2 + \left(\frac{2\pi\gamma}{T}\right)^2\right\}}} \sin\frac{2\pi(t_0 - t_1)}{T} + Ce^{-\frac{R}{\gamma}t_0},$$

$$0 = \frac{A}{R} - \frac{E}{\sqrt{\left\{R^2 + \left(\frac{2\pi\gamma}{T}\right)^2\right\}}} \sin\frac{2\pi(t_0 - t_1)}{T}$$

$$+ Ce^{-\frac{R}{\gamma}t_0} \cdot e^{-\frac{RT}{2\gamma}},$$

equations to determine t_0 and C.

Eliminating C,

$$\frac{RE}{A\sqrt{\left\{R^2 + \left(\frac{2\pi\gamma}{T}\right)^2\right\}}} \cdot \sin\frac{2\pi(t_0 - t_1)}{T} = -\tanh\frac{RT}{4\gamma}.$$

Having obtained t_o, C is given by equation

$$\frac{2\,A}{R} = C e^{-\frac{R}{\gamma}t_0}\left(1 + e^{-\frac{R}{2\gamma}T}\right).$$

This gives the complete solution of the problem.

A case of special importance is that in which R is small; let us therefore consider the case $R = 0;$ the solution then is

$$\gamma\,x = -\frac{T}{2\,\pi}\,E\cos\frac{2\,\pi\,t}{T} - A\,t + C.$$

In the same way as before,

$$E\cos\frac{2\,\pi\,t_0}{T} = \frac{A\,\pi}{2},$$

$$C = A\left(t_0 + \frac{T}{4}\right).$$

The limiting case to which the solution applies is given by $x' = 0$ when $t = t_0 + \frac{T}{2}.$

$$E\sin\frac{2\,\pi\,t_0}{T} = A,$$

whence $\qquad\qquad E^2 = A^2\left(1 + \frac{\pi^2}{4}\right),$

or $\qquad\qquad A = E \times 0.538.$

Roughly, we may say that, in order that the current may not cease for a finite time, E must be at least double of A;

A will of course depend upon the length of the arc.   The work done in the arc will be proportional to the arithmetical mean value of the current taken without regard to sign.   This is of course quite a different thing from the mean current as measured by an electro-dynamometer. Let us examine what error is caused by estimating the work done in the arc as equal to the current measured by the dynamometer multiplied by the mean difference of potential.

The actual work done per second

$$= \frac{2\,A}{T} \int_{t_0}^{t_0 + \frac{T}{2}} x\,d\,t$$

$$= \frac{2\,A}{\pi\,\gamma} \cdot \frac{T}{2\,\pi} \cdot \sqrt{E^2 - \frac{\pi^2 A^2}{4}}.$$

The mean square of the current as measured by the electro-dynamometer is

$$\frac{2}{T} \int_{t_0}^{t_0 + \frac{T}{2}} x^2\,d\,t = \overline{\frac{T}{2\,\pi\,\gamma}}^2 \left\{ \frac{E^2}{2} - 2\,A^2 \right\} + \frac{A^2\,T^2}{48\,\gamma^2},$$

and the work done by this current is apparently the square root of the above expression multiplied by A.   It is easy to see that this is greater in all cases than the work done, but it is worth while to examine the extent of the error. If we treated the arc as an ordinary resistance, we should assume work per second

$$= \frac{A}{\gamma} \sqrt{\left(\frac{T}{2\,\pi}\right)^2 \left(\frac{E^2}{2} - 2\,A^2\right) + \frac{A^2\,T^2}{48}}.$$

Taking a fairly practical case, assume $A = \dfrac{2}{5} E$; we have actual work per second

$$= \frac{A' T}{\gamma} \cdot \frac{1}{\pi^3} \sqrt{\frac{25}{4} - \frac{\pi^3}{4}}$$

$$= \frac{A'}{\gamma} T \frac{\sqrt{15}}{20} \text{ nearly;}$$

work done estimated by electro-dynamometer

$$= \frac{A' T}{\gamma} \sqrt{\frac{1}{40} \left(\tfrac{1}{2} \frac{25}{4} - 2 \right) + \frac{1}{48}} \text{ nearly,}$$

$$= \frac{A' T}{\gamma} \frac{1}{20} \sqrt{\frac{235}{12}} \,,$$

or nearly $\frac{1}{8}$ part too much. This will suffice to show that the matter is not a mere theoretical refinement. Another erroneous method of estimating the power developed in an arc is to replace by a resistance and adjust this resistance till the current as measured by an electro-dynamometer is the same as with the arc, and assume that the work done in the resistance is the same as the work done in the arc.

Returning to the expression

$$\frac{2 A}{\pi \gamma} \cdot \frac{T}{2 \pi} \sqrt{E^3 - \frac{\pi^3 A^3}{4}} \,,$$

we may inquire, given $T A$ and the dimensions of the machine, how ought it to be wound or its coils connected that

most work may be done in the arc. If the number of convolutions be varied, E will vary as the convolutions, γ as
their square; therefore $\gamma \propto E^2$; we are therefore to deter-

mine E so that $\dfrac{1}{E^2} \sqrt{E^2 - \dfrac{\pi^2 A^2}{4}}$ is a maximum which

occurs when $E = \pi A$. When the resistance of the circuit
is taken into account, this result will be modified. It
suffices to prove that it is desirable that the potential of
the machine should be materially in excess of that required
to maintain the arc.

V.* In all that precedes it is assumed, not only that γ is
constant, but that the copper conductor of the armature is
the only conductor moving in the field. If there be iron
cores in the armature, we shall approximate to the effect
by regarding such cores as a second conducting circuit.
Slightly changing the notation, let L be coefficient of self
induction of the copper circuit, N coefficient of self induction of the iron circuit and R' its resistance, I' the magnetic induction of the field magnets upon the iron circuit,
and M the coefficient of mutual induction of the two circuits, y the current in the iron. The equations of motion
are obtained from the expression for the energy, viz.,

$$\tfrac{1}{2}\{L x^2 + 2 M x y + N y^2 - 2 I x - 2 I' y\},$$

and are

$$L x' + M y' + R x = \frac{d I}{d t} = \frac{2 \pi A}{T} \cos \frac{2 \pi t}{T},$$

$$M x' + N y' + R' y = \frac{d I'}{d t} = \frac{2 \pi B}{T} \cos \frac{2 \pi t}{T},$$

_____

* Vide also "Encyclopædia Britannica," article "Lighting."

for in general the iron cores and the copper conductor are symmetrically arranged. Assume

$$x = a \sin \frac{2 \pi t}{T} = b \cos \frac{2 \pi t}{T},$$

$$y = a' \sin \frac{2 \pi t}{T} + b' \cos \frac{2 \pi t}{T},$$

and substitute in the equations of motion; we have the following four equations to determine the constants a, b, a', b':—

$$a \frac{2 \pi L}{T} + a' \frac{2 \pi M}{T} + R\, b = \frac{2 \pi A}{T},$$

or

$$
\left.
\begin{aligned}
a L + a' M + b \frac{T R}{2 \pi} &= A, \\[2ex]
b L + b' M - a \frac{T R}{2 \pi} &= O, \\[2ex]
a M + a' N + b' \frac{T R'}{2 \pi} &= B. \\[2ex]
b M + b' N - a' \frac{T R'}{2 \pi} &= O.
\end{aligned}
\right\}
$$

and

These equations contain the solution of the problem, but are too cumbersome to be worth while solving generally;

we will, however, prove the statements made in the lecture before the Civil Engineers.

1. Compare short circuit and open circuit, that is, $R = O$ very nearly, and $R = \alpha'$. In the former case we find that work done in the iron is diminished, and if $B = \dfrac{A\,M}{L}$ we have the paradoxical result that there are no currents induced in the iron of the cores and no work is required to drive the machine. This, of course, can never actually occur, because R can never absolutely vanish. It suffices to show, however, that the current in the copper circuit may diminish the whole power required to drive the machine to an amount less than the power required to drive the machine on open circuit.

2. The other statement related to the effect of the currents in the iron upon the currents produced in the copper circuit. Assume that the effect is a small one, for a first approximation. Neglect it, that is, treat the currents in the iron and the currents in the copper as independent of each other, and then see how each would disturb the other.

The first approximation then is

$$\left\{
\begin{aligned}
a &= \frac{A\,L}{L^{2} + \dfrac{T^{2}R^{2}}{4\,\pi^{2}}}; & a' &= \frac{B\,N}{N^{2} + \dfrac{T^{2}R'^{2}}{4\,\pi^{2}}}; \\[3em]
b &= \frac{A\,\dfrac{T\,R}{2\,\pi}}{L^{2} + \dfrac{T^{2}R^{2}}{4\,\pi^{2}}}, & b' &= \frac{B\,\dfrac{T\,R'}{2\,\pi}}{N^{2} + \dfrac{T^{2}R'^{2}}{4\,\pi^{2}}}.
\end{aligned}
\right.$$

If we substitute these in the general equations as corrections, we have

$$
\begin{cases}
a\,L + b\,\dfrac{T\,R}{2\,\pi} = A - \dfrac{B\,N\,M}{N' + \dfrac{T'\,R''}{4\,\pi^2}} \\[4ex]
-\,a\,\dfrac{T\,R}{2\,\pi} + b\,L = -\,\dfrac{B\,\dfrac{T\,R'}{2\,\pi}\,M}{N' + \dfrac{T'\,R''}{4\,\pi^2}},
\end{cases}
$$

which shows that the disturbing effect of each circuit upon the other is to diminish the apparent electromotive force, but to accelerate its phase.

VI. A very similar problem is that of secondary generators or induction coils, whether used for the conversion of high potentials to low, or the reverse. To treat it generally, taking the magnetization of the iron cores, which are always used, as a non-linear function of the currents in the coils, would be a matter of much difficulty; we therefore assume, as is usual, that the coefficients of induction are constants, noting in passing that this is not strictly the fact, though it is very nearly the fact, when the cores are not saturated and when the lines of magnetic induction pass through non-magnetic space.

Let, then, R, r be the resistances of the primary and secondary circuits;

L coefficient of self induction of the primary;

N coefficient of self induction of the secondary;

M coefficient of mutual induction of the two circuits;

x and y the currents in the two circuits at time t;

X the electromotive force applied in the primary circuit by an alternate current dynamo machine or otherwise;

the equations of motion will be

$$\left. \begin{array}{l} L\,x' + M\,y' + R\,x = X, \\ M\,x' + N\,y' + \ r\,y = 0. \end{array} \right\}$$

Various assumptions may be made as to X, but that most likely to be adopted in the practical work of secondary generators is that X is kept so adjusted that

$$x = A \cos n\,t \text{ where } n = \frac{2\,\pi}{T},$$

and to inquire how X will depend on the resistances

$$N\,y' + r\,y = n\,A\,M \sin n\,t,$$

$$y = \frac{n\,A\,M}{n^2\,N^2 + r^2}\,(-\,n\,N \cos n\,t + r \sin n\,t),$$

$$X = A\left[\left\{ -\,n\,L + \frac{n^2\,N\,M^2}{n^2\,N^2 + r^2}\right\}\,\sin n\,t \right.$$

$$\left. +\,\left\{\frac{M^2\,n^2\,r}{n^2\,N^2 + r^2} + R\right\}\,\cos n\,t\right].$$

As in the case of the dynamo machine, the work done in the secondary circuit is greatest when $r = n\,N$. The expression for X serves to show that when the secondary is short circuited a *lower* electromotive force of the generating circuit is required than when it is on open circuit. In induction coils the electrostatic capacity of the coils themselves has important effects. An illustration of the effect of electrostatic induction is found in the old-fashioned Ruhmkorff coils. These were not wound symmetrically, but in such wise that one end of the secondary coil was on the whole towards the inside, the other towards the outside of the bobbin. In such coils a spark to earth may be obtained from the outside end, but not from the inside. The reason is that the outer convolutions have smaller electrostatic capacity than the inner ones. The terminals may be made to give equal sparks by the simple expedient of laying a piece of tinfoil around the whole coil and connecting it to earth.

VII. Some time ago Dr. Muirhead told me that he had observed that the effect of an alternate current machine could be increased by connecting it to a condenser. This is not difficult to explain: it is a case of resonance analogous to those which are so familiar in the theory of sound and in many other branches of physics.

Take the simplest case, though some others are almost as easy to treat. Imagine an alternate current machine with its terminals connected to a condenser; it is required to find the amplitude of oscillation of potential between the two sides of the condenser. Let $R\ \gamma$ be the resistance and self induction of the machine, $E \sin \dfrac{2\,\pi\,t}{T}$ its electromotive force, C the capacity of the condenser, V the difference of potential sought, and x the current in the machine; then

$$C V' = x,$$

and

$$\gamma x' + R x = E \sin \frac{2 \pi t}{T} - V;$$

whence

$$\gamma x'' + R x' = \frac{2 \pi}{T} E \cos \frac{2 \pi t}{T} - \frac{x}{C},$$

$$x = \frac{\left\{ 1 - C \gamma \frac{2 \pi}{T} \right\} \cos \frac{2 \pi t}{T} + R C \frac{2 \pi}{T} \sin \frac{2 \pi t}{T}}{\left\{ 1 - C \gamma \left(\frac{2 \pi}{T} \right)^2 \right\}^2 + R^2 C^2 \left(\frac{2 \pi}{T} \right)^2} \cdot \frac{2 \pi E C}{T},$$

$$V = \frac{\left\{ 1 - C \gamma \left(\frac{2 \pi}{T} \right)^2 \right\} \sin \frac{2 \pi t}{T} - R C \frac{2 \pi}{T} \cos \frac{2 \pi t}{T}}{\left\{ 1 - C \gamma \left(\frac{2 \pi}{T} \right)^2 \right\}^2 + R^2 C^2 \left(\frac{2 \pi}{T} \right)^2} \cdot E;$$

amplitude of V is therefore

$$= \frac{1}{\sqrt{ \left\{ 1 - C \gamma \left(\frac{2 \pi}{T} \right)^2 \right\}^2 + R^2 C^2 \left(\frac{2 \pi}{T} \right)^2 }} \cdot E.$$

Now suppose $E = 100$ volts, the machine would light up an incandescent lamp of about 69 volts. Let $T = \frac{1}{200}$ second,

$C = 100$ microfarads, and $\dfrac{2\,\pi\,\gamma}{T} = 8$ ohms, and $R = \frac{1}{10}$ ohm, all figures which could be practically realized; we have amplitude of $V = 80\ E$ roughly, or the apparent electromotive force would be increased eighty fold.

We now return to the principal subject of the present communication. Some attempts have been made to verify the proposition that two alternate current machines can be advantageously connected parallel, but, I believe, till recently without success. I had no convenient opportunity for testing the point myself till last summer, when I had two machines of De Meritens, intended for the lighthouse of Tino, in my hands. I have made no determinations of the constants of these machines, but between three and four years ago I thoroughly tested a pair of similar machines now in use at a lighthouse in New South Wales. Each machine has five rings of sixteen sections, and forty permanent magnets. The resistance of the whole machine as connected for lighthouse work (a single arc) was 0.0313, its electromotive force (E) when running 830 revolutions per minute, 95 volts and $\left(\dfrac{2\,\pi\,\gamma}{T}\right) = 0.044$ ohm. It was further remarked that the loss of power was least with a maximum load, as is shown in the following table:—

Power applied as measured in belt,	3.1	4.8	5.6	6.5	5.4
Electric power developed	0.7	3.4	4.3	5.7	3.4
Mean current in ampères	7.7	38.6	51.7	73.6	151

This result illustrates well the conclusion arrived at in Problem V. above.

Last summer the two machines for Tino were driven

from the same countershaft by link bands, at a speed of
850 to 900 revolutions per minute; the pulleys on the
countershaft were sensibly equal in diameter, but those on
the machines differed by rather more than a millimètre,
one being 300, the other 299 mms. in diameter (about);
thus the two machines had not when unconnected exactly
the same speed. The pulleys have since been equalized.
The bands were of course put on as slack as practicable,
but no special appliance for adjusting the tightness of the
bands was used. The experiment succeeded perfectly at
the very first attempt. The two machines, being at rest,
were coupled in series with a pilot incandescent lamp
across the terminals; the two bands were then simultaneously thrown on: for some seconds the machines almost
pulled up the engine. As the speed began to increase, the
lamp lit up intermittently, but in a few seconds more the
machines dropped into step together, and the pilot lamp
lit up to full brightness and became perfectly steady and
remained so. An arc lamp was then introduced, and a perfectly steady current of over 200 ampères drawn off without disturbing the harmony. The arc lamp being removed,
a Siemens electro-dynamometer was introduced between
the machines, and it was found that the current passing
was only 18 ampères, whereas, if the machines had been in
phase to send the current in the same direction, it would
have been more than ten times as great. On throwing off
the two bands simultaneously, the machines continued to
run by their own momentum, with retarded velocity. It
was observed that the current, instead of diminishing from
diminished electromotive force, steadily increased to about
50 ampères, owing to the diminished electrical control be-

tween the machines, and then dropped off to zero as the machines stopped. Professor Adams will, I hope, give an account of experiments he has tried with me, and on other occasions, at the South Foreland. With De Meritens' machines, I regard coupling two or more machines parallel as practically the best way of obtaining exceptionally great currents when required in a lighthouse for penetrating a thick atmosphere.

AN UNNOTICED DANGER IN CERTAIN APPA-RATUS FOR DISTRIBUTION OF ELECTRICITY.

MANY plans have been proposed, and several have been to a greater or less extent practically used, for combining the advantage of economy arising from a high potential in the conductors which convey the electric current from the place where it is generated with the advantages of a low potential at the various points where the electricity is used. A low potential is necessary where the electricity is used; partly because the lamps, whether arc or incandescent, each require a low potential, and partly because a high potential may easily become dangerous to life. Among the plans which have been tried for locally transforming a supply of high potential to a lower and safer, the most promising is by the use of secondary generators or induction coils. It has been proved that this method can be used with great economy of electric power and with convenience; under proper construction of the induction coils it may also be perfectly safe. It is, however, easy and very natural so to construct them that they shall be good in all other respects but that of safety to life—that they shall introduce an un-expected risk to those using the supply.

In a distribution of electricity by secondary generators, an alternating current is led in succession through the primary coils of a series of induction coils, one for each

group or system of lamps. The lamps connect the two
terminals of the secondary coil of the induction coils. It
is easy to so construct the induction coils that the difference of potential between the terminals of the secondary
coils may be any suitable number of volts, such as 50 or
100; while the potential of the primary circuit, as measured between the terminals of the dynamo machine, may
be very great, e.g., 2,000 or 3,000 volts. If the electromagnetic action between the primary and secondary coils, on
which the useful effect of the arrangement depends, were
the only action, the supply would be perfectly safe to the

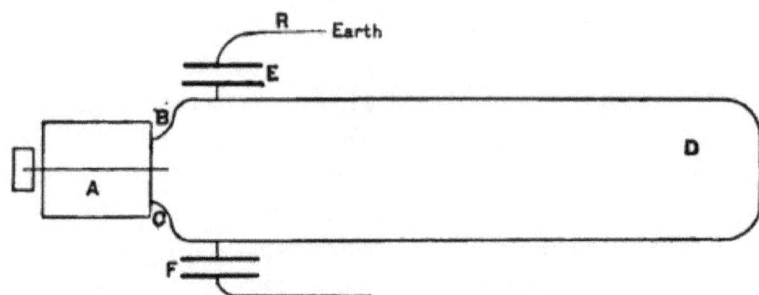

FIG. 55.

user so long as apparatus with which he could not interfere
was in proper order. But the electromagnetic action is
not the only one. Theoretically speaking, every induction
coil is also a condenser, and the primary coil acts electro-
statically as well as electromagnetically upon the secondary
coil. This electrostatic action may easily become dangerous if the secondary generator is so constructed that its
electrostatic capacity, regarded as a condenser, is other
than a very small quantity.

Imagine an alternate current dynamo machine, A, Fig. 55, its terminals, B, C, connected by a continuous conductor, $B D C$, on which may be resistances, self induction coils, secondary generators, or any other appliances: at any point is a condenser, E, one coating of which is connected to the conductor, or may indeed be part of it; the other is connected to earth through a resistance, R. Let K be the capacity of the condenser, V the potential at time t of the earth coating of the condenser, U the potential of the other coating, x the current in resistance R to the condenser from the earth, being taken as positive, and the earth potential as zero. We have

$$x = \frac{V}{R}, \ K(U^{\cdot} - V^{\cdot}) = x;$$

whence, since

$$U = A \sin 2\, \pi\, n\, t,$$

where A is a constant depending on the circumstances of the dynamo circuit as well as the electromotive force of the machine, and n is the reciprocal of the periodic time of the machine, we have

$$K R \overset{\cdot}{x} + x = 2\, \pi\, n\, K\, A \cos 2\, \pi\, n\, t,$$

$$x = \frac{2\, \pi\, n\, K\, A}{(K R\, 2\, \pi\, n)^2 + 1}\{-2\, \pi\, n\, K R \sin 2\, \pi\, n\, t + \cos 2\, \pi\, n\, t\},$$

$$\text{mean square of } x = \frac{2\, \pi\, n\, K}{\sqrt{(K R\, 2\, \pi\, n)^2 + 1}} \cdot \text{mean square of } A.$$

Let us now consider the actual values likely to occur in practice. Let the condenser E be a secondary generator;

let the resistance R be that of some person touching some part of the secondary circuit, and also making contact to earth with some other part of the body; n may be anything from 100 to 250, say 150; K will depend on the construction of the secondary generator—it may be as high as 0.3 microfarad or even more, but there would be no difficulty even in large instruments in keeping it down to one hundredth of this or less. The mean square of A will depend on the circumstances of other parts of the circuit; it might very easily be as great, or very nearly as great, as the mean difference of potential between the terminals of the machine if the primary circuit were to earth at C. Suppose, however, that the circuit $B\ D\ C$ is symmetrical, that E is at one end, and that another person of the same resistance as the person at E is touching the secondary circuit of the secondary generator F at the other end of the circuit. In that case, if 2,400 be difference of potential of the machine, mean square of A will be 1,200; in which case we have, taking R as 2,000 ohms,

mean square of

$$x = \frac{2\,\pi \times 150 \times 0.3 \times 10^{-6}}{\sqrt{(2\,\pi \times 150 \times 0.3 \times 10^{-6} \times 2,000)^2 + 1}} \times 1,200$$

$$= \text{about } 0.3 \text{ ampère.}$$

Experiments are still wanting to show what current may be considered as certain to kill a man, but it is very doubtful whether any man could stand 0.3 ampère for a sensible length of time. It is probable that if the two persons both 'took firm hold of the secondary conductors of E and F, both would be killed. If the person at F be replaced by an accidental dead earth on the secondary circuit of F, the person at E would experience a greater current than 0.3 ampère.

It follows from the preceding consideration that secondary generators of large electrostatic capacity are essentially dangerous, even though the insulation of the primary circuit and of the primary coils from the secondary coils is perfect. The moral is—for the constructor, Take care that the secondary generators have not a large electrostatic capacity, say not more than 0.03 microfarad, better less than $\frac{1}{100}$ microfarad; for the inspector, Test the system for safety. The test is very easy. Place a secondary generator of greatest capacity at one end of the line and connect its secondary circuit to earth through any instrument suitable for measuring alternate currents under one ampère; put the other end of the primary to earth; the reading of the current measuring instrument should not exceed such a current as it may be demonstrated a man can endure with safety.

INDUCTION COILS OR TRANSFORMERS.

THE transformers considered are those having a continuous iron magnetic circuit of uniform section.*

Let A be area of section of the core;

> m and n the number of convolutions of the primary and secondary coils, respectively;
>
> R, r, and ρ their resistances, ρ being the resistance of the secondary external to the transformer;
>
> x and y currents in the two coils;
>
> a induction per square centimetre;
>
> α the magnetic force;
>
> l the length of the magnetic circuit;
>
> $E = B \sin 2 \pi (t/T)$, the difference of potentials between the extremities of the primary;
>
> T being the periodic time.

We have

$$4 \pi (m x + n y) = l \alpha; \tag{1}$$

$$E = R x - m A a; \tag{2}$$

$$0 = (r + \rho) y - n A a. \tag{3}$$

---

* For a discussion of transformers in which there is a considerable gap in the magnetic circuit, see Ferraris, Torino, *Accad. Sci. Mem.*, vol. 37, 1885; also chapter on the "Theory of Alternating Currents," in this volume.

From (2) and (3),

$$n E = n R x - m (r + \rho) y. \qquad (4)$$

Substituting from (1),

$$x\{n^2 R + m^2 (r + \rho)\} = n^2 E + (l \alpha/4 \pi) m (r + \rho); \quad (5)$$

$$y\{n^2 R + m^2 (r + \rho)\} = - n m E + (l \alpha/4 \pi) n R; \quad (6)$$

$$A a = - \frac{(r + \rho) m E}{n^2 R + m^2 (r + \rho)} + \frac{l \alpha R (r + \rho)}{4 \pi \{n^2 R + m^2 (r + \rho)\}}. \quad (7)$$

We may now advantageously make a first approximation. Neglect $l \alpha$ in comparison with $4 \pi m x$, that is, assume the permeability to be very large; we have

$$A a = - \frac{(r + \rho) m B \sin (2 \pi t/T)}{n^2 R + m^2 (r + \rho)}; \qquad (8)$$

$$A a = \frac{(r + \rho) m B \cos (2 \pi t/T)}{\{n^2 R + m^2 (r + \rho)\} . 2 \pi t/T}. \qquad (9)$$

For practical purposes these equations are really sufficient.

We see first that the transformer transforms the potential in the ratio n/m, and adds to the external resistance of the secondary circuit ρ a resistance $(n^2 R/m^2) + r$. This at once gives us the variation of potential caused by varying the number of lamps used. The phase of the secondary current is exactly opposite to that of the primary.

In designing a transformer it is particularly necessary to take note of equation (9), for the assumption is that a is limited so that $l \alpha$ may be neglected. The greatest value of a is $B/\{(2\pi/T)\,m\,A\}$, and this must not exceed a chosen value. We observe that B varies as the number of reversals of the primary current per unit of time.

But this first approximation, though enough for practical work, gives no account of what happens when transformers are worked so that the iron is nearly saturated, or how energy is wasted in the iron core by the continual reversal of its magnetism. The amount of such waste is easily

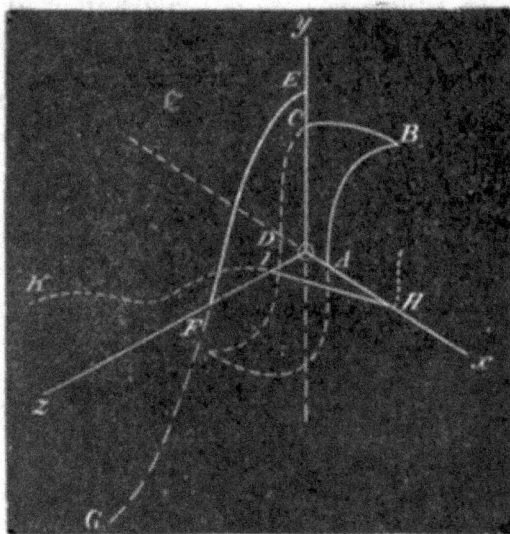

FIG. 56.

estimated from Ewing's results when the extreme value of a is known, but it is more instructive to proceed to a second approximation, and see how the magnetic properties of the iron affect the value and phase of x and y. We shall, as a second approximation, substitute in equations (5), (6), (7) values of α deduced from the value of a furnished by the first approximation in equation (9).

In the accompanying diagram, Fig. 56, Ox represents α, Oy represents a, and Oz the time t.

The curves $A\,B\,C\,D$ represent the relations of a and α. $E\,F\,G$ the induction a as a function of the time, and $H\,I\,K$ the deduced relation between a and t. We may substitute the values of α obtained from this curve in equations (5) and (6), and so obtain the values of x and y to a higher degree of approximation. If the values of α were expressed

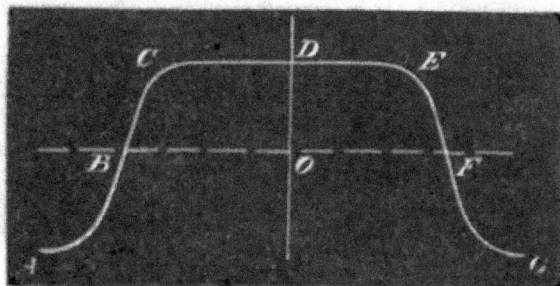

Fig. 57.

by Fourier's theorem in terms of the time, we should find that the action of the iron core introduced into the expression for x and y, in addition to a term in $\cos (2\,\pi\,t/T)$ which would occur if a and α were proportional, terms in $\sin (2\,\pi\,t/T)$ and terms in sines and cosines of multiples of $2\,\pi\,t/T$. It is through the term in $\sin (2\,\pi\,t/T)$ that the loss of energy by hysteresis comes in.

A particular case, in which to stay at a first approximation would be very misleading, is worthy of note. Let an attempt be made to ascertain the highest possible values of

a by using upon a transformer a very large primary current
and measuring the consequent mean square of potential in
the secondary circuit by means of an electrometer, by the
heating of a conductor, or other such device. The value
of *a* will be related to the time somewhat as indicated by
A B C D E F G in Fig. 57; for simplicity assume it to be
as in Fig. 58; the resulting relations of potential in the

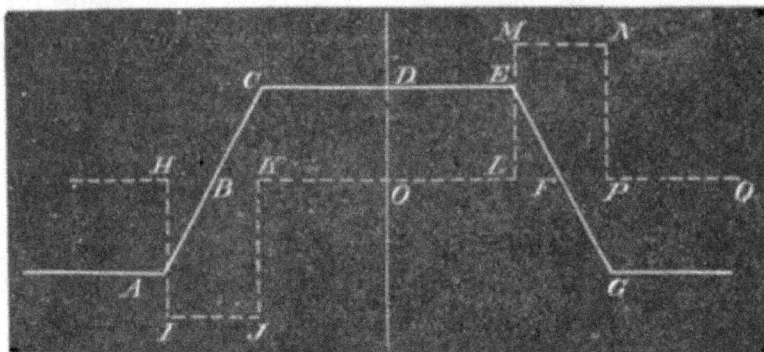

Fɪɢ. 58.

secondary and the time will be indicated by the dotted line
H I J K O L M N P Q. The mean square observed will be
proportional to *M L . √L P;* but *M L . L P* is proportional
to *E L,* hence the potential observed will vary inversely as
√L P, even though the maximum induction remain constant. If, then, the maximum induction be deduced on
the assumption that the induction is a simple harmonic
function of the time, results may readily be obtained vastly
in excess of the truth.

REPORT TO THE WESTINGHOUSE COMPANY OF THE TEST OF TWO 6,500-WATT WESTINGHOUSE TRANSFORMERS.

BEFORE giving any of the results of the tests I have made with your transformers, it will be well to explain the methods of experiment adopted. The instantaneous value at any epoch in the period of the difference of potential between any two points of a circuit in which the potential difference is varied periodically is made effective on the measuring instrument by means of a rotating contact maker attached to the shaft of the alternate current generator. This contact maker was constructed for the King's College laboratory by Messrs. Siemens Brothers. It makes contact once in each revolution for a period of about three quarters of a degree, and breaks it for the rest of the revolution. It is entirely insulated, and so can be connected to any part of the circuit. The position of the contact can be varied, and the variation be read off on a graduated circle of 13½ inches diameter divided into degrees, and by estimation the variation can be read to one tenth of a degree. The two points between which it is desired to measure a potential difference are connected through the contact maker to a condenser and a quadrant electrometer, as shown in Fig. 59, in which A and B are the points, the potential difference of which at a stated epoch is to be measured, C the revolving contact maker, D the reversing switch of the electrometer, E the condenser, of which the capacity can be varied, F the quadrant electrometer. It is evident that the quadrant electrometer will give a reading proportional to the potential difference of A and B, when C makes contact. If there were no leakage, it would at

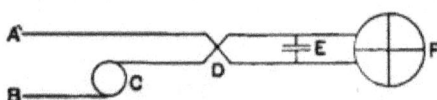

Fig. 59.

once give this potential. It is to obviate the effect of leakage that the condenser is introduced, and the amount of the effect was determined by varying the condenser thus: When the condenser had capacity 1, 0.5, and 0.2 microfarads, the readings of the electrometer for a given potential difference of an alternating current at the position in the period of maximum electromotive force were 138, 136, and 132, respectively. The rate of loss of potential will be proportional to the reciprocal of the capacity, whence we infer that the true reading, if insulation were perfect, would be $139\frac{1}{2}$, and hence the readings are always corrected by adding one per cent. When the potential difference was too great for the electrometer it was reduced in any desired ratio by two considerable resistances introduced between the points to be measured in the usual way (Fig. 60). The potential difference may, of course, be measured in other ways. An ordinary voltmeter may be

placed between *A* and *B*, in which case it must be standardized with the contact breaker in circuit; and it will depend
for its constant on the duration of the contact, which may
vary. Further, it gives, not the difference of potential

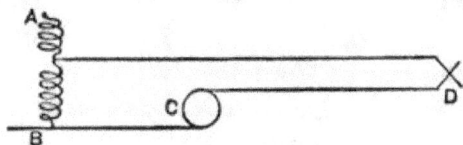

Fig. 60.

at any definite epoch, but the mean difference for the
whole time of the contact. The condenser may be used
and its potential be measured by discharge through a galvanometer: this is open to the objection that if there be
any leakage, the result will depend on the time at which
the contact is broken by the condenser key in relation to
the time at which it was made by the revolving contact
maker. Lastly, a Clark cell may be used, by a method
which Major Cardew pointed out to me (Fig. 61), the re-

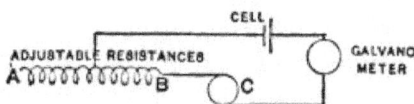

Fig. 61.

sistance being adjusted till there is no deflection. This
is open to the same objection as the first, namely, that it
gives the mean of the potential differences which occur
during the contact. By making use of the first-mentioned
method we have the means of measuring accurately any
potential difference at any epoch of the period, and of
knowing the epoch.

For these experiments two transformers intended to be identical were available, each transforming between 2,400 and 100 volts. It was most convenient on account of the resistance available to couple these transformers up from 100 to 2,400 in the first or No. 1 transformer, then down from 2,400 to 100 in the second or No. 2 transformer, and to take up the energy from the second in a non-inductive resistance. The arrangement is shown in Fig. 62.

The obvious way in returning the efficiency of the combination would be to measure, at various epochs of half a period, the potential differences of the terminals of the

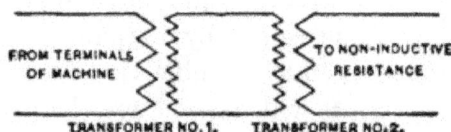

FIG. 62.

machine and the current passing in the No. 1 transformer; in like manner, at the same epochs, to measure either the potential differences or the current passing to the non-inductive resistance, thence to deduce the power supplied to the first transformer and taken from the second. This would be open to certain objections: we are comparing two nearly equal magnitudes and desire their ratio; the ratio will be afflicted with the full error arising from an error in the determination of either magnitude, and such errors may be material, as the observations are not simultaneous, and conditions may change between one series of observations and another.

These objections are avoided by the method adopted. The current from No. 2 is observed at certain epochs, the

difference of current between No. 2 and No. 1, and the
difference of potential difference of No. 2 and No. 1 at the
same epoch. These give the currents and potentials of
No. 1 at the same epochs as the corresponding determinations of No. 2, and the difference will only be afflicted with
the proportion of error of those differences. For example,
suppose the efficiency of the combination were 90 per cent.,
and the possible error of determination of power 1 per
cent., our result might be anything from 38 per cent. to 92
per cent. if made in the obvious way, but if made by differences the maximum loss would be 10.1 per cent., and
the possible least determination of the efficiency would be
89.8 per cent. The method is essentially similar to the

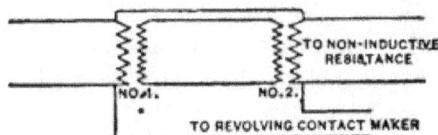

FIG. 63.

method I described* and subsequently used for testing
dynamos. The measurements for difference of potential
differences are made as in Fig. 63. For current differences
(Fig. 64), where G is a known small non-inductive resist-

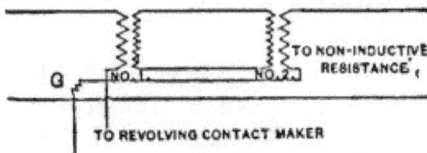

FIG. 64.

ance, the two currents will, of course, slightly disturb each
other, but this is readily allowed for in the calculations.

_____

* *Phil. Trans.*, 1886, page 347.

Another method would be to couple **them** as in Fig.
65, G_1 and G_2 being equal non-inductive resistances.
This **arrangement is** quite free from disturbance, but re-
quires two resistances adjusted to exact equality. A single
transformer can be tested in the same way, though in this

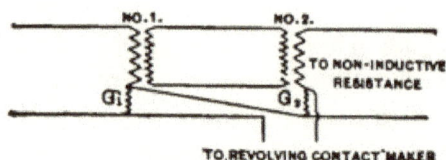

FIG. 65.

case reliance must be placed upon resistances to reduce the
current of the low potential coil, and to reduce the potential of the high potential coil in the ratio of the number of
windings in the two coils.

The current was throughout generated by a Siemens
alternator with 12 magnets, run at a speed between 830
and 840 revolutions per minute, which gives a frequency
of 5,000 per minute, or 83 to 84 per second.

The first experiment tried * was with the two transformers coupled, but with No. 2 transformer on open circuit, or
on nearly open circuit, for a high resistance for purposes
of measurement was interposed between the terminals of
the low resistance coil of No. 2 transformer. The actual

---

* So far as I know, the first discussion of endless magnetic circuit transformers, based on the actual properties of the material, is in a note by myself (*Proc.
Roy. Soc.*, vol. XLII., and *The Electrician*, vol. XVIII., p. 421.) Definite results
were obtained by methods generally similar to those now used by Prof. Ryan
(*The Electrical World*, Dec. 28, 1889). The theory of transformers is well set
forth by Prof. Fleming (*The Electrician*, April 22 and 29, 1892).

results are given in Table IX., and are expressed in Fig. 66. Tables X., XI., and XII. give the results for half power,

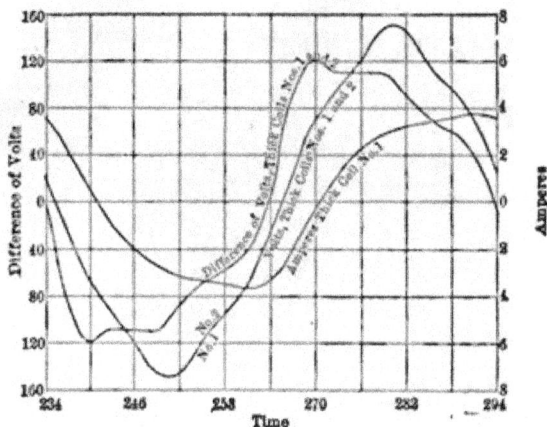

FIG. 66.

TABLE IX.

Leads of Exploring Brush on Divided Circle.	Current No. 1. Thick Coils. Ampères.	Potential No. 2. Thick Coils.		Potential No. 1. Thick Coils.		Square of Volts. $\sqrt{\text{mean}^2}$ = 101.9.	Watts supplied to No. 1.
		Volts.	Square of Volts. $\sqrt{\text{mean}^2}$ = 101.1	P. D. Sec Nos. 1 and 2. Volts.	Volts.		
267	− 2.2	+ 25.4	645	+ 0.9	+ 26.3	692	− 57.9
270	− 0.3	+ 70.2	4,928	+ 1.2	+ 71.4	5,098	− 21.4
273	+ 1.1	+ 95.3	9,082	+ 1.1	+ 96.4	9,292	+ 106.0
276	+ 2.1	+120.4	14,496	+ 1.1	+121.5	14,761	+ 255.1
279	+ 2.8	+147.7	21,816	+ 1.1	+148.8	22,140	+ 416.6
282	+ 3.2	+147.2	21,668	+ 0.9	+148.1	21,935	+ 473.9
285	+ 3.4	+119.8	14,351	+ 0.7	+120.5	14,520	+ 409.7
288	+ 3.5	+ 97.8	9,565	+ 0.6	+ 98.4	9,683	+ 344.4
291	+ 3.7	+ 71.3	5,084	+ 0.4	+ 71.7	5,140	+ 260.3
294	+ 3.5	+ 26.0	676	+ 0.3	+ 25.97	674	+ 90.9
			102,311			103,935	2,282.6

nearly full power, and full power, and the sets of curves of Figs. 67, 68 and 69 give the results of the table. In these tables the first column gives the position of the contact brush in degrees, so that 60 on this scale corresponds with a complete cycle. Three degrees are thus $\dfrac{1}{83.3 \times 20}$ of a second. The second column of Table IX. is the current in the thick coil of No. 1 transformer, as determined by the difference of potential at the two ends of a noninductive resistance in which the current passes. The third column is the potential difference of No. 2 transformer, a direct determination. The fourth column is solely for the purpose of determining the square root of the mean of the squares of the third column. This column is a direct determination of the difference of potential of No. 1 and No. 2, obtained in the manner explained with reference to Fig. 63.

The sixth column is the deduced potential difference of the terminals of the thick wire of No. 1 transformer, being the sum of the third and fifth columns. The seventh column, like the fourth, is merely for the purpose of determining the square root of the mean of the squares of column six, while the eighth gives the rate at which power is given out or received by the pair of transformers.

If the transformers had been exactly equal, the potentials for the two given by Table IX. would have been equal, though they would have differed a little in phase owing to the lines of magnetic induction which pass through the non-magnetic space between the two coils of the transformer.* The difference shows that No. 1 transformer has

---

* Prof. Perry has already pointed out that the effect of such an induction cannot be entirely neglected, even in endless circuit transformers.

a ratio of transformation slightly greater than No. 2.   If
we correct the potential of either No. 1 or No. 2, there

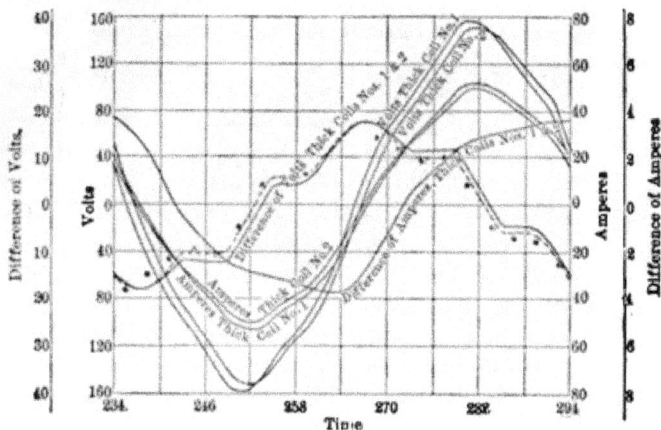

FIG. 67.

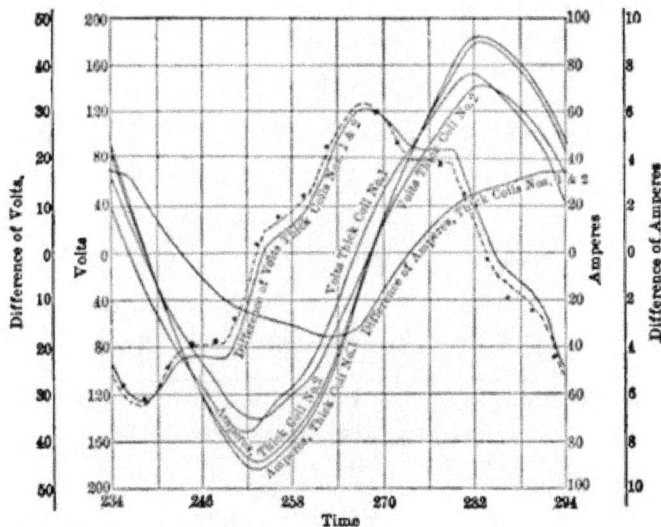

FIG. 68.

still remains a difference between them, but this difference
will be greatest about when the potentials are *nil*. This

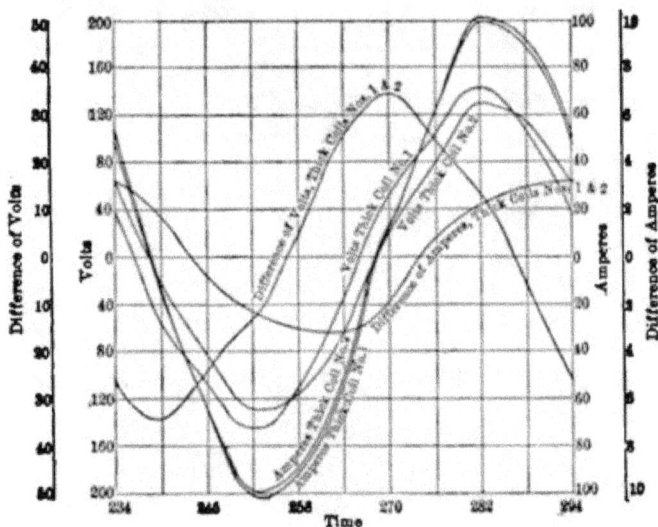

Fig. 69.

is due to the lost induction just referred to. In order to
check the conclusion that the two transformers are not

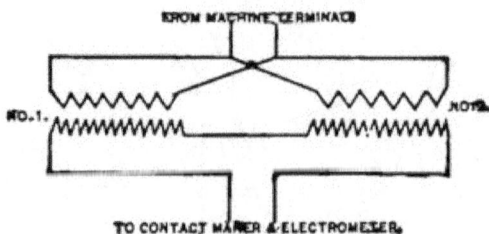

Fig. 70

precisely equal, they were directly compared, as in Fig. 70.
The transformers were coupled parallel, as in Fig. 70, and

the difference of potential of the two high potential coils was measured: the value of its square root of mean square was 12.5 volts, the potential of the transformer being 2,400. This does not necessarily imply that the potentials of the two transformers differ by one half per cent.; it may be largely due to a difference of phase between the two.

The current supplied to the No. 1 transformer is to be accounted for by the currents necessary to magnetize the

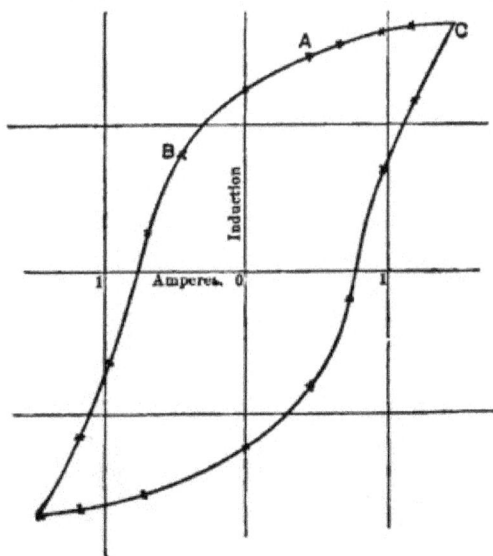

FIG. 71.

two transformers, and by the local currents in their cores. To ascertain the former, the curve of magnetization of one of the transformers was determined by the ballistic galvanometer for nearly the same induction as in Table IX., the changes of current supplied by a battery being made by a reversing switch, or by suddenly introducing resistance

into the primary circuit, and the consequent changes of
induction being measured by the galvanometer. The
tardiness of change of current in the transformer due to
its self induction was sufficiently reduced by using many
cells and a considerable resistance. The results are shown
in Fig. 71 for a single transformer. In this curve the
abscissæ are the currents in the thin coil of the trans-

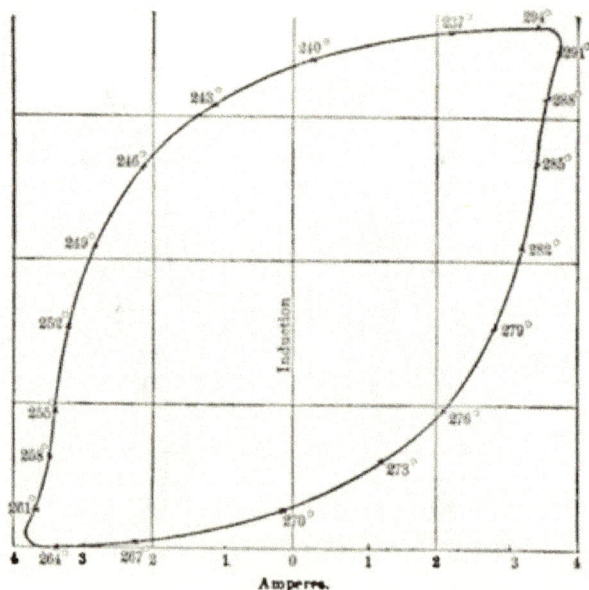

FIG. 72.

former, divided by 24 to reduce it to the same effect as it
would have had if it had been in the thick coil. The
ordinates are the inductions as measured by the kick on
the galvanometer, but reduced to a scale to make them
directly comparable with the volts when the transformer is
used with an alternating current. These results are not

given in absolute units. The procedure to determine
points on the curve was: first, pass the maximum current
corresponding to the point C; next, suddenly diminish the
current by inserting suitable resistance in the thin coil
circuit, and observe the kick—the drop of ordinate from
C to A corresponds to the kick, and the abscissa of A is
the current after it has been reduced; next, reverse the
current, and observe the kick—the kick corresponds to the
further drop of ordinates from A to B. In this manner a
series of points are determined on the curve. Fig. 72
shows the relation between induction and magnetizing
current for the pair of transformers, as deduced from the
experiments with alternating currents set forth in Table IX.
The ordinates in this curve are the area of the curve of
potentials of Fig. 66, for the ordinates of this latter curve
are the rates at which the induction is changing, while the
abscissæ are the currents in the thick wire at corresponding
times. The points marked • in Fig. 73 give the remainder
after deducting the magnetizing current as estimates in
Fig. 71 from the currents of Fig. 72—that is to say, Fig.
71 is corrected first for the small difference in maximum
induction; then, corresponding to any induction, the current is taken from the curve, it is doubled, as there is only
one transformer, and the result is deduced from the corresponding current of Fig. 72. The differences are, the
magnetizing current equivalent and opposite in effect to
the local currents in the cores. If the local currents were
equivalent to a current in a single secondary circuit, the
points • of Fig. 73 ought to have had the form of the full
line of Fig. 73, drawn through the points +, in which the
abscissæ are proportional to the potential difference, and

the ordinates to the induction. Returning to Table IX., we
find that the fall of potential difference on open circuit in
the whole combination is 0.8 volt, and that the loss of

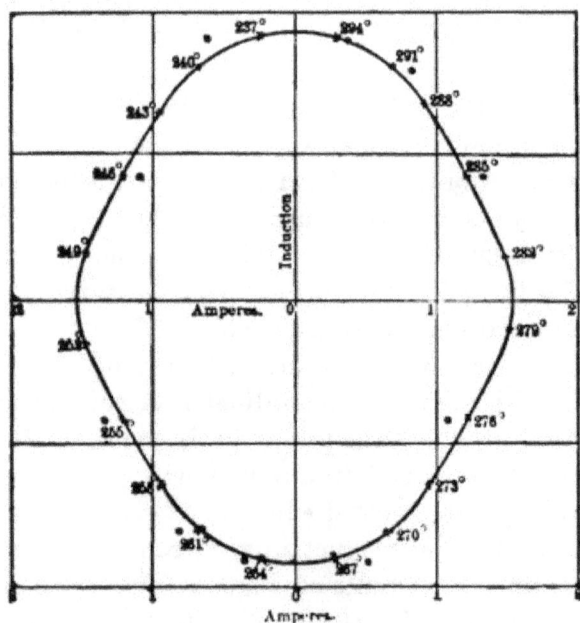

Fig. 73.

power in magnetizing the cores and in local currents is
222.86 watts, that is, a loss for each transformer of 114.13
watts. The total loss of 228 watts may be divided into 126
watts accounted for by hysteresis and 102 watts due to
local currents.

Referring now to Table XI. and Fig. 68, the earlier columns explain themselves, but a word is necessary about
the last six columns. The watts supplied to No. 1 are
simply the products at each time of the volts at its termi-

nals and the ampères passing through, similarly to the watts given out by No. 2. We see first that the efficiency of the whole combination with this load is 93.73 per cent., and hence the efficiency of one transformer, if the losses in the two are equal, may be taken as 96.9 per cent. The fall of potential in the whole combination is 6.1 volts, but the fall with no load is 0.8 volt; hence the variation due to the load with constant potential on the thin coil of No. 1 is 5.3 volts, or, if the fall of potential in the two transformers were equal, which it is not, for a single transformer 2.65 volts. Assuming that the transformers are equal, the power lost in resistance would be expected to be the mean of mean current × the difference of potential difference, or 215.4 watts. It is, in fact, 150 watts, as given by multiplying the square of currents by resistances. But the transformers are not exactly equal, and there is the waste magnetic field, both of which will have a small effect on the distribution of loss between the two classes of loss,—viz., that by hysteresis and local currents, and that by resistance,—but none upon the gross efficiency.

The other tables, X. and XII., are arranged in exactly the same way as Table XI., but the number of observations on Table XII. is insufficient to bring out all the peculiarities of the transformers.

It has already been stated that, if the loss of potential due to load in the two transformers be equal, it will amount to 2.65 per cent. The following experiment was tried to ascertain if this loss was equal: The transformers were coupled in series as before. The mean potential difference of the thick wire was measured by Thomson's multicellular, and of the thin wire by Thomson's electrostatic volt-

meter. The mean of a considerable number of experiments
is given in the following table, the load being the same as
in Table XI., and the results being corrected to the same
potential of the thin wire:—

Number.	Full Load.		Open Circuit.	
	Thomson's Multicellular.	Thomson's Electrostatic.	Thomson's Multicellular.	Thomson's Electrostatic.
1	2,380	99.8	2,380	97.0
2	2,380	94.2	2,380	96.2

This shows that of a total drop of 4.8 volts, 2.8 volts oc-
curred in No. 1, and 2 volts in No. 2. There is no doubt
of the fact that the drop is greater in No. 1 than in No. 2,
which is connected with the waste field between the two
coils. Of course these transformers are intended to work
exactly as No. 2 is working, in which case the drop from
no load to nearly full load, as shown by this experiment, is
2.0 volts. The way in which this waste field causes inequality of drop of potential in the two transformers, coupled
as in my experiments, is well worthy of careful consideration. The waste field is proportional to the current in the
transformers, or, better, to the mean of the two currents in
ampère turns. The electromotive force due to this waste
field will be proportional to the rate of change of the current. If the current were expressed by a simple harmonic
curve, the electromotive force due to the waste field would
also be a simple harmonic curve differing in phase by $\frac{N}{2}$.
The curve of potentials is roughly in the same phase as the

curve of current. Let A be the amplitude of potential difference of No. 2 transformer, B be the amplitude of difference of potential difference in No. 2, or the potential difference of the thin wire divided by 24. $2b$ will be very nearly the amplitude of difference between the thick wires of Nos. 1 and 2. The ratios of potentials in No. 1 and No. 2 will then be

$$\frac{\sqrt{a^2 + - 4b^2}}{\sqrt{a^2 + b^2}} \text{ and } \frac{\sqrt{a^2 + b^2}}{a}, \text{ or } 1 + \frac{3}{2}\frac{b^2}{a^2} \text{ and } 1 + \frac{b^2}{2a^2},$$

or the drop in the first from this cause is three times as great as in the second transformer. We shall return to the waste field immediately. Putting aside harmonic curves, and returning to the facts as they are, the following table gives: first, half the difference of potential difference

Half Difference of Potential Difference.	Volts of High Potential Coil Divided by 24.	Squares of Volts.
15.6	−5.0	25.0
14 8	41.3	1,706.0
11.3	74.3	5,520.0
10.8	102.6	10,530.0
11.1	131.3	17,240.0
5.5	147.1	21,640.0
− 1.1	137.0	18,770.0
− 3.1	119.3	14,230.0
− 5.9	95.1	9,040.0
−12.1	53.4	2,850.0

Square root of mean square = 100.8.

taken from Table XI., that is, at each instant the drop of potential in No. 2; secondly, the volts of the thin wire of No. 2 reduced for number of convolutions—this is of course the mean of potential difference between 1 and 2; lastly, the squares of these volts. From this we see a mean square 100.8 showing the drop in No. 2 to be 2.6 volts out of a total drop of 6.1, and the remainder 2.5, the drop in No. 1. Diminishing these results by 0.4, the half of 0.8, the fall observed with no load, the actual losses from no load to nearly full load will be 2.2 and 3.1.

Turn now to the last column of Table XI. This gives the difference of potential differences corrected for the loss of volts by resistance. It is shown dotted on Fig. 68; this curve presents one or two peculiar features. It should be possible to infer the form of this curve from the curve of current. The rates at which the mean current is changing are as follows:

268½	271½	274½	277½	280½	283½	286½	289½	292½	....
30.7	24.2	19.2	18.5	13.9	1.7	−9.3	−12.7	−21.7	−28.1

—which happens to come to a scale which can be at once plotted. The points marked • are the points of the curve corresponding with the above rates. The agreement of the points with the curve is remarkably close. This exhibits very completely the effect of waste magnetic field in this transformer.

For half power, as taken from Table X., the rates are as follows:

268½	271½	274½	277½	280½	283½	286½	289½	292½	295½
14.6	11.5	9.4	10.6	4.4	−4.8	−7.2	−7.5	−13.6	−17.6

—and in the same way in Fig. 67 the dotted curve represents the difference of electromotive force corrected for resistance, and the points correspond with the above rates.

Fig. 74 gives the efficiencies for the combined transformers in terms of the load. This curve is the hyperbole:

$$\text{Efficiency} = 100 \cdot \frac{X - (A + B \, X \, C \, X')}{X},$$

where A = 228, the loss by hysteresis;

    B = 0.005, and mainly depends upon the waste field;

    C = 0.0000035, and is mainly the loss by resistance;

    X = load in watts.

To sum up, I find that the efficiency of the transformer at full load would be 96.9 per cent.; at half load, 96 per

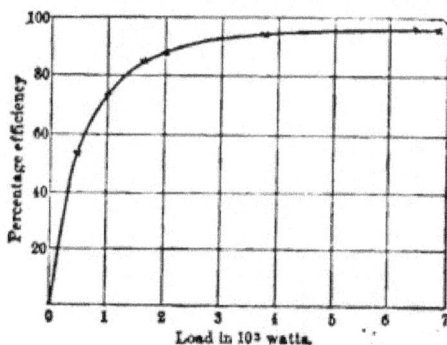

Fig. 74.

cent.; and at quarter load, over 92 per cent. The magnetizing current of the transformer amounts to 114 watts, or 1.75 per cent. The drop of potential from no load to full load is between 2 per cent. and 2.2 per cent.

In conclusion, I wish to express my thanks to Mr. Wilson, of King's College; this gentleman carried out the experiments under my direction, and made nearly all the numerical calculations and drew most of the curves for me.

TABLE X.

Leads of Exploring Brush on Divided Circle.	Current No. 2. Thick Coil.			Current No. 1. Thick Coil.			Current Difference. Thick Coils, Nos. 1 and 2.		Mean of Currents in Nos. 1 and 2. Thick Coils.	Potential No. 2. Thick Coil.	
	Observed Deflection.	Corrected Deflection.	Ampères.	Observed Deflection.	Corrected Deflection.	Volts.	Volts.	Ampères.	Ampères.	Volts.	Square of Volts. $\sqrt{\text{mean}^2} = 104.$
267	2.5	2.5	+ 0.6	189.5	189.0	+ 3.7	− 2.9	− 9.8	− 0.8	+ 5.6	31
270	65.0	65.0	+ 16.0	77.0	77.0	+ 1.5	− 1.2	+ 14.8	+ 15.4	+ 51.9	2,694
273	106.5	106.5	+ 26.7	28.0	28.0	− 0.6	+ 9.4	+ 27.1	+ 26.9	+ 88.6	6,969
276	145.0	145.0	+ 35.6	99.0	99.0	− 1.9	+ 1.5	+ 37.1	+ 36.8	+ 110.3	12,165
279	186.0	186.0	+ 45.7	152.5	152.5	− 3.0	+ 2.4	+ 48.1	+ 46.9	+ 141.1	19,910
282	205.0	205.2	+ 49.9	186.0	186.0	− 3.7	+ 2.9	+ 52.8	+ 51.8	+ 153.6	23,592
285	184.0	184.0	+ 45.2	203.0	201.2	− 4.0	+ 8.1	+ 48.8	+ 46.7	+ 138.5	19,181
288	154.0	154.0	+ 37.8	220.0	218.0	− 4.3	+ 2.4	+ 41.2	+ 39.5	+ 114.9	13,209
291	123.0	123.0	+ 30.2	238.5	235.5	− 4.6	+ 8.6	+ 38.8	+ 32.0	+ 90.6	8,208
294	67.7	67.7	+ 16.6	237.0	234.0	− 4.6	+ 8.6	+ 20.2	+ 18.4	+ 47.7	2,275
											108,247

[This table is continued on the next page.]

TABLE X.—*Continued.*

Leads of Exploring Brush on Divided Circle.	Potential Difference. Thick Coils, Nos. 1 and 2.			Potential Difference No. 1. Thick Coil.		Mean of Potential in Nos. 1 and 2, Thick Coils.	Watts supplied to No. 1.	Watts given out by No. 2.	Loss by Resistance and probably Waste Field.		Lo-s by Hysteresis and Local Currents.	Difference of Potential due to Waste Field.
	Observed Deflection.	Corrected Deflection.	Volts.	Volts.	Square of Volts. √mean² = 107.4.				Resistance.	Losses unaccounted for, probably Waste Field.		
267	315.0	308.0	+ 18.2	+ 23.8	566	+ 14.7	− 55	+ 3	0.02	− 14.62	− 42.6	+ 18.2
270	278.0	273.0	+ 16.1	+ 68.0	4,634	+ 60.0	+ 1,006	+ 830	8.7	+ 239.2	− 72.0	+ 15.5
273	196.5	195.0	+ 11.5	+ 95.1	9,044	+ 89.3	+ 2,577	+ 2,332	25.6	+ 282.7	+ 35.7	+ 10.5
276	99.0	197.2	+ 11.6	+ 121.9	14,860	+ 116.1	+ 4,523	+ 3,925	48.5	+ 372.5	+ 174.1	+ 10.3
279	203.0	201.2	+ 11.9	+ 153.0	23,405	+ 147.1	+ 7,359	+ 6,448	80.9	+ 477.2	+ 353.0	+ 10.2
282	61.0	61.0	+ 3.6	+ 157.2	24,710	+ 155.4	+ 8,560	+ 7,665	96.8	+ 87.9	+ 451.6	+ 1.7
285	73.0	73.0	− 4.3	+ 134.2	18,009	+ 136.3	+ 6,482	+ 6,290	90.3	− 281.3	+ 422.5	− 6.0
288	75.0	75.0	− 4.4	+ 110.5	12,210	+ 112.7	+ 4,552	+ 4,343	57.5	− 231.3	+ 383.0	− 5.9
291	126.5	126.5	− 7.5	+ 83.1	6,904	+ 86.9	+ 2,908	+ 2,736	51.7	− 277.7	+ 312.8	− 8.7
294	252.0	248.5	− 14.7	+ 33.0	1,089	+ 40.3	+ 606	+ 792	12.5	− 283.0	+ 145.1	− 15.4
					115,424		38,518 52.2%	35,226 52.2%	449.5	371.6	2,162.4	

298.3

298.5

UNIVERSITY OF CALIFORNIA LIBRARY

TABLE II.

Leads of Exploring Brush on Divided Circle.	Current No. 2, Thick Coil. Ampères.	Current No. 1, Thick Coil.			Mean of Current in Nos. 1 and 2, Thick Coils.	Potential No. 2, Thick Coil.		Potential Difference No. 1, Thick Coil.				Mean of Potential in Nos. 1 and 2, Thick Coils.	Watts supplied to No. 1.	Watts given out by No. 2.	Loss by Resistance and probably Waste Field.			Loss by Hysteresis and Local Currents.	Difference of Potential due to Waste Field.
		Ampères.	Current Difference. Secondaries. Nos. 1 and 2, Ampères.			Volts.	Square Volts. $\sqrt{mean^2}=98.2$.	Potential Difference. Thick Coils, Nos. 1 and 2, Volts.	Volts.	Square Volts. $\sqrt{mean^2}+104.3$.					Resistance.	Losses unaccounted for, probably Waste Field.			

TABLE XII.

Leads of Exploring Brush on Divided Coil.	Current No. 2. Thick Coil.			Current Difference. Thick Coils. Nos. 1 and 2.				Current No. 1. Thick Coil.	Mean of Current in Nos. 1 and 2. Thick Coils.	Potential No. 2. Thick Coil.	
	Observed Deflection.	Corrected Deflection.	Amperes.	Observed Deflection.	Corrected Deflection.	Volts.	Amperes.	Amperes.		Volts.	Square of Volts. $\sqrt{mean^2} = 90.9$.
270	59	59	+ 14.3	107	107	+ 2.1	− 1.7	+ 12.6	+ 13.4	+ 21.3	454
276	256	252	+ 61.0	57	57	− 1.1	+ 0.9	+ 61.9	+ 61.4	+ 80.7	6,512
282	427	410	+ 99.3	147	147	− 2.9	+ 2.3	+ 101.6	+ 100.4	+ 130.2	16,950
288	381	309	+ 89.3	188	187	− 3.7	+ 2.9	+ 92.2	+ 90.8	+ 116.1	13,480
294	207	205	+ 49.6	211	209	− 4.1	+ 8.2	+ 52.8	+ 56.2	+ 62.4	3,893
											41,289

[This table is continued on the next page.]

TABLE XII.—Continued.

Leads of Exploring Brush on Divided Coil.	Potential No. 1. Thick Coil.					Mean of Potential in Nos. 1 and 2. Thick Coils.	Watts supplied to No. 1.	Watts given out by No. 2.	Losses by Resistance and probably Waste Field.		Loss by Hysteresis and Local Currents.	Difference of Potential due to Waste Field.
	Potential Difference. Thick Coils, Nos. 1 and 2.			Volts.	Square of Volts. $\sqrt{\text{mean}^2} = 98.4$.				Resistance.	Losses unaccounted for, probably Waste Field.		
	Observed Deflection.	Corrected Deflection.	Volts.									
270	855	845	+ 33.9	+ 55.2	3,047	38.8	696	304	6.8	+ 447.5	− 65.1	+ 38.4
276	256	252	+ 24.8	+ 105.5	11,130	93.1	6,590	4,928	142.8	+ 1,380.2	+ 88.8	+ 52.5
282	142	142	+ 14.0	− 144.2	20,730	127.2	14,650	12,928	889.0	+ 1,028.5	+ 815.5	+ 10.2
288	59	59	− 5.8	+ 110.3	12,165	113.2	10,170	10,367	312.3	− 889.0	+ 889.8	− 9.2
294	266	262	− 25.8	+ 30.6	1,940	49.5	1,893	3,095	99.8	− 1,430.3	+ 158.4	− 27.7
					48,472		33,978	31,617	943.2	591.9	880.9	
							98.05%		472.2	471.2		

THEORY OF THE ALTERNATE CURRENT DYNAMO.

ACCORDING to the accepted theory of the alternate current dynamo, the equation of electric current in the armature is $\gamma\, y + R\, y = $ periodic function of t, where γ is a constant coefficient of self induction. This equation is not strictly true, inasmuch as γ is not in general constant,* but it is a most useful approximation. My present purpose is to indicate how the values of γ and of the periodic function representing the electromotive force can be calculated in a machine of given configuration.

To fix ideas, we will suppose the machine considered to have its magnet cores arranged parallel to the axis of rotation, that the cores are of uniform section, also that the armature bobbins have iron cores, so that we regard all the lines of induction as passing either through an armature coil or else between adjacent poles entirely outside the armature. The sketch, Fig. 75, shows a development of the machine considered. The iron is supposed to be so arranged that the currents induced therein may be neglected. We further suppose for simplicity that the line integral of magnetic force within the armature core may be neglected.

---

* See chapter on the " Theory of Alternating Currents" in this volume.

Let A_1 be the effective area of the space between the pole piece and armature core when the cores are in line, l_1 the distance from iron to iron.

Let A_2 be the section of magnet core, l_2 the effective length of a pair of magnet limbs, so that l_2 may be re-

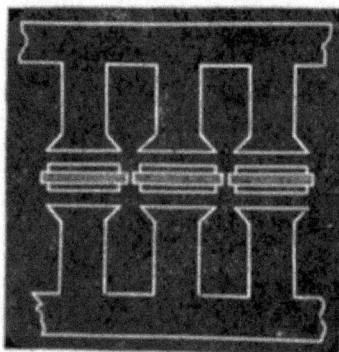

Fig. 75.

garded as the length of the lines of force as measured from one pole face to the next.

Let m be the number of convolutions in a pair of magnet limbs, and

n the convolutions in one armature section;

T the periodic time.

The time is measured from an epoch when the armature coil we shall consider is in a symmetrical position in a field which we shall regard as positive.

x and y are the currents in the magnet and armature coils, the positive direction being that which produces the positive field at time zero.

At time t the armature coil considered has area A_1',

$$= b_0 + b_1 \cos (2 \pi t / T) + b_2 \cos (4 \pi t / T) + \text{etc.,}$$

in a positive field; and area A_1'',

$$= b_0 - b_1 \cos(2\pi t/T) + b_2 \cos(4\pi t/T) - \text{etc.},$$

in a negative field, where

$$b_0 + b_1 + b_2 + \ldots = A_1,$$

and

$$b_0 - b_1 + b_2 + \ldots = 0.$$

The coefficients b_0, b_1, etc., are deducible by Fourier's theorem from a drawing of the machine under consideration.

Let I be the total induction in the magnet core, and let, at time t, I be distributed into I' through A_1', I'' through A_1'' and I''' as a waste field to the neighboring poles.

The line integral of magnetic force from the pole to either adjacent pole is I'''/k, where k is a constant.

We have first to determine I', I'', I''', in terms of x and y.

Take the line integral of magnetic force in three ways through the magnets, and respectively through area A_1', through area A_1'', and across between the adjacent poles—

$$l_2 f\left(\frac{I}{A_2}\right) + 2 l_1 \frac{I'}{A_1'} = 4\pi m x + 4\pi n y,$$

$$l_2 f\left(\frac{I}{A_2}\right) + 2 l_1 \frac{I''}{A_1''} = 4\pi m x - 4\pi n y,$$

$$l_2 f\left(\frac{I}{A_2}\right) + \frac{I'''}{k} = 4\pi m x;$$

whence

$$I' + I'' + I''' = I$$

$$= \left(\frac{A_1' + A_1''}{2\,l_1} + k\right)\left(4\,\pi\,m\,x - l_2 f\left(\frac{I}{A_2}\right)\right)$$

$$+ \frac{A_1' - A''}{2\,l_1} \cdot 4\,\pi\,n\,y.$$

When t, x and y are given, this would suffice to determine I by means of the known properties of the material of the magnets as represented by the function f. We will, however, consider two extreme cases between which other cases will lie.

First. Suppose that the intensity of induction in the magnet cores is small, so that $l_2 f (I/A_2)$ may be neglected, the iron being very far from saturation. We have

$$I' - I'' = \frac{4\,\pi}{2\,l_1} \{m\,(A_1' - A_1'')\,x + n\,(A_1' + A_1'')\,y\}$$

$$= \frac{4\,\pi}{l_1} \left\{ m \left(b_1 \cos \frac{2\,\pi\,t}{T} + b_3 \cos \frac{6\,\pi\,t}{T} + \ldots\right) x \right.$$

$$\left. + n \left(b_0 + b_2 \cos \frac{4\,\pi\,t}{T} + \ldots\right) y \right\}.$$

We see that the coefficient of self induction y in general contains terms in $\cos (4\,\pi\,t/T)$.

Second. In actual work it would be nearer the truth to

suppose that the magnetizing current x is so great that the induction I may be regarded as constant, and the quantity $l_2 f(I/A_2)$ as considerable. But as small changes in I imply very great changes in $l_2 f(I/A_2)$, its value cannot be regarded as known. We have then

$$2 l_1 \frac{I'}{A_1'} = \frac{2 l_1}{A_1' + A_1'' + 2 k l_1}$$

$$\left(I - \frac{A_1' - A_1''}{2 l_1} \cdot 4 \pi m y \right) + 4 \pi m y,$$

$$2 l_1 \frac{I''}{A_1''} = \frac{2 l_1}{A_1' + A_1'' + 2 k l_1}$$

$$\left(I - \frac{A_1' - A_1''}{2 l_1} \cdot 4 \pi m y \right) - 4 \pi n y;$$

whence

$$I' - I'' = \frac{(A_1' - A_1'') I}{A_1' + A_1'' + 2 k l_1}$$

$$- \left\{ \frac{(A_1' - A_1'')^2}{A_1' + A_1'' + 2 k l_1} - (A_1' + A_1'') \right\} \frac{4 \pi n y}{2 l_1}$$

$$= \frac{(A_1' - A_1'') I}{A_1' + A_1'' + 2 k l_1}$$

$$+ \frac{4 A_1' A_1'' + 2 k l_1 (A_1' + A_1'')}{A_1' + A_1'' + 2 k l_1} \cdot \frac{4 \pi n y}{2 l_1}.$$

For illustration, consider the simplest possible case: let $b_0 = b_1 = \frac{1}{2}A_1$, and $b_2 = b_3 = \ldots = 0$, and let $2\,k\,l_1$ be negligible; we have

$$I' - I'' = I \cos \frac{2\,\pi\,t}{T} + A_1 \sin^2 \frac{2\,\pi\,t}{T} \cdot \frac{4\,\pi\,n\,y}{2\,l_1},$$

and the equation of current will be

$$R\,y = n \left\{ \frac{2\,\pi}{T} I \sin \frac{2\,\pi\,t}{T} - \frac{2\,\pi\,n\,A_1}{l_1} \frac{d}{d\,t} \left(\sin^2 \frac{2\,\pi\,t}{T} \cdot y \right) \right\},$$

instead of the simple and familiar linear equation.

THE ELECTRIC LIGHTHOUSES OF MACQUARIE AND OF TINO.

The subject of the use of the electric light in light-houses was fully discussed at the Institution in 1879, when papers by Sir James Douglass, M. Inst. C.E., and by Mr. James T. Chance, Asso. Inst. C.E., were read.*

The subject has been further elaborately examined by Mr. E. Allard,† and more recently in practical experiments made at the South Foreland, exhaustively reported on by a committee of the Trinity House.‡ The justification of the present communication is that, at the lighthouses of Macquarie and of Tino, the optical apparatus is on a larger scale than has hitherto been used for the electric arc in lighthouses, and presents certain novel features in the details of construction. Further, as regards the electrical apparatus, tests were made upon the machinery for Macquarie when it was in the hands of Messrs. Chance Brothers & Company, which still possess some value, although five years old; and, in the case of Tino, the machines are practically worked together in a manner not previously used otherwise than by way of experiment.

---

* Minutes of *Proceedings of the Inst. C.E.*, vol. LVII., pp. 77 and 168.

† " Mémoire sur les Phares Électriques," 1881.

‡ " Report into the relative merits of Electricity, Gas, and Oil as Lighthouse Illuminants." Parts 1 and 2. PP. 1885.

In the case of both lighthouses, Messrs. Chance Brothers & Company, of Birmingham, entered into a contract for the supply of all the apparatus required, including engines, machines, conductors, lamps, optical apparatus, and lanterns; and Sir James Douglass, engineer-in-chief of the Trinity House, acted as inspecting engineer to the respective colonial and foreign governments.

As these two lighthouses present many features in common, it may be most convenient to give a full description of the earlier lighthouse, and then limit the description of Tino to those points in which it differs from Macquarie.

MACQUARIE.

This lighthouse is situated on South Head, near Sydney, the precise position being shown in a copy from the chart, Fig. 76. A lighthouse was first placed at this important landfall in 1817. The focal plane is 346 feet above the sea, and the distance of the sea horizon is therefore 21.6 nautical miles, and the range about 27 nautical miles for an observer 15 feet above the sea.

Optical Apparatus.—The light is a revolving one, giving a single flash of eight seconds' duration every minute. On account of the considerable altitude of the lighthouse, it was necessary to secure that a substantial quantity of light should be directed to the nearer sea; but it was also essential, on account of the exceptional power of the apparatus, that this dipping light should only be a small fraction of that sent to the horizon, otherwise its effect would be excessively dazzling. Many years ago Mr.

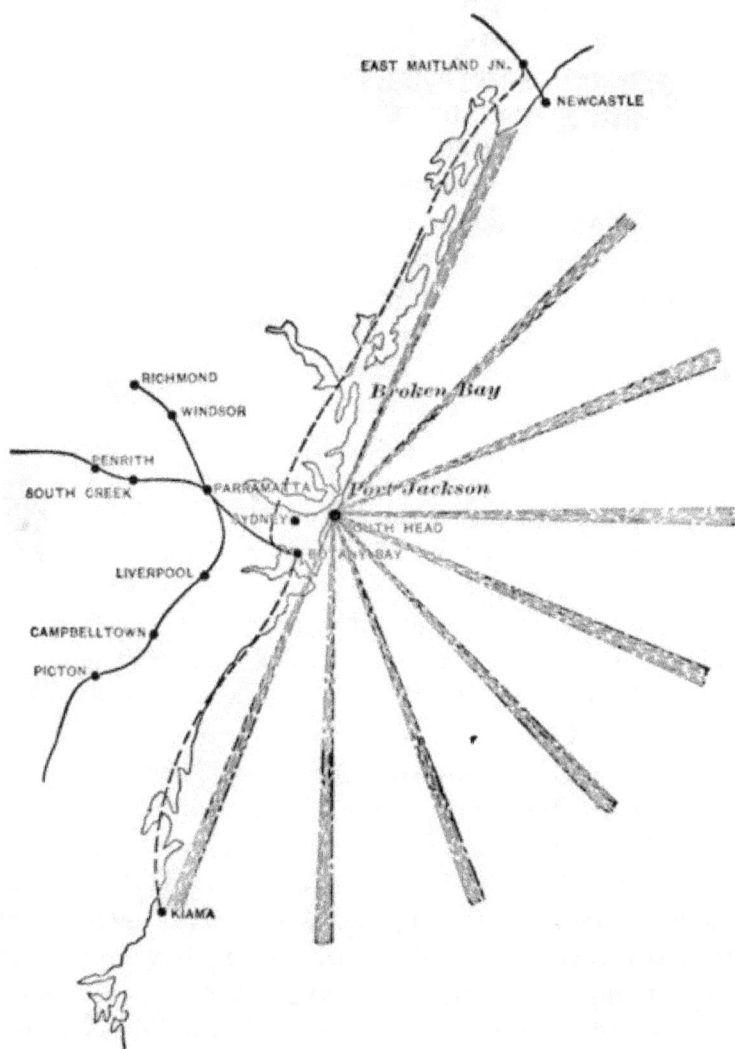

FIG. 76.

James T. Chance urged that it was not wise to make use
of very small apparatus for the electric arc, because a
larger apparatus renders it possible for the optical engineer
to effect with greater precision the distribution of light
which is most desirable, and because any trifling error
which may occur in the position of the electric arc has,
with the larger apparatus, a less marked effect on the light
as seen from the sea.   In the lighthouses of Souter Point,
the South Foreland, and the Lizard, the third order apparatus of 500 millimetre focal length was adopted.
Optically, the larger the apparatus used the better, but
there might be some question whether, on purely optical
grounds, the advantage of going beyond the third order is
sufficient to justify the additional expense; but in the case
of a revolving apparatus the third order is a very incon-
venient size for the service of the lamp—it is too large to
be conveniently served from the outside, and too small to
admit the attendant within it with comfort.   With the
large currents, which are now easily obtained and are
likely to be used in lighthouses, a first or second order
apparatus has the further advantage that it is less liable to
injury from particles thrown off from the heated carbons.
In the case of Macquarie, it was decided to adopt an apparatus of the first order, 920 millimetre focal length; it
was further decided that the optical apparatus should produce its condensing effect by means of a single agent—that
is to say, the vertical straight prisms which were used in
Souter Point and other revolving electric lighthouses
should be dispensed with.   The condensation and distribution of light necessary may be obtained by means of
a single agent, with apparatus such as has been pro-

posed by Mr. Alan Brebner, Jr., Asso. M. Inst. C.E.;*
but this construction is open to the objections that it
is somewhat costly, and that it increases the length of
the path of the rays through the glass, and consequent
absorption. A practically better plan is to adopt forms
not differing very greatly from those introduced by
Fresnel; to specially arrange them for the purpose in
hand, and to accept certain consequent minute deviations
from a mathematically accurate solution for the sake of
advantages of greater importance when all the actual conditions are taken into account. Fig. 77 shows the optical
apparatus in vertical section: the upper and the lower
totally reflecting prisms are, as is usual in revolving lights,
forms of revolutions about a horizontal axis; they direct
the light incident upon them to the horizon and the distant sea from 10' above the horizon to 30' below; they are
specially adjusted to distribute the light in azimuth over
the arc of 3° necessary for a proper duration of flash.

The refracting portion of the apparatus has the profile
so calculated that the central lens and the three rings
next to the lens above and below direct their light to the
horizon without vertical divergence, except what is due to
the size of the arc. The light for the nearer sea is obtained
from the remaining ten lens segments, Nos. 5 to 9 inclusive, above and below the centre, counting the centre as
No. 1, the distribution being according to the following
table, in which the first column gives the denomination of
the elements of the lens in accordance with the numbers
marked upon the section; the second, the angle between

---

* Minutes of *Proceedings of the Inst. C.E.*, vol. LXX., p. 386.

I.	II.			III.		
	°	′	″	°	′	″
9 top	—10	0		2	30	59
8 "	2	30	59	5	8	52
7 "	—10	0		2	37	30
6 "	—10	0		1	30	0
5 "	—10	0		1	0	0
5 bottom	—10	0		1	0	0
6 "	—10	0		1	30	0
7 "	1	30	0	3	44	27
8 "	3	44	27	5	50	41
9 "	5	50	41	7	46	57

the direction of the sea horizon and the ray emerging from the upper limit of the element; the third, the angle between the direction of the sea horizon and the ray from the lower limit of the element, the negative sign denoting that the emerging ray is above the horizon. This practice of appropriating certain elements of the apparatus to different distances on the sea was first introduced by Mr. James T. Chance, in the lights of the South Foreland exhibited in January, 1872.

The ray, dipping at an angle of 7° 46′ 57″ below the horizon, will strike the sea at ½ mile, while 5° 8′ 52″ corresponds to ¾ mile, 2° 37′ 30″ to 1¼ mile, 1° 30′ to 2 miles, 1° to 2½ miles, and 30′ to about 4 miles. Thus the direct light begins at about ½ mile from the lighthouse. From ½ mile to ¾ mile the sea receives light from one element of the apparatus, from ¾ to 1¼ mile from two elements, from 1¼ mile to 2 miles from three elements, from 2 to 2½ miles from four elements, and beyond 2½ miles from six elements; the upper and lower totally reflecting prisms come in aid at about 5 miles. The main power of the apparatus is hardly attained till a distance of 8 or 10 miles. Fig. 78 is a sectional plan of the apparatus by a horizontal

FIG. 77.

FIG. 79.

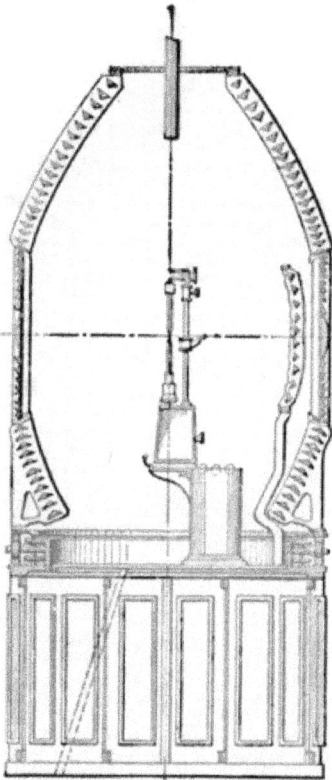

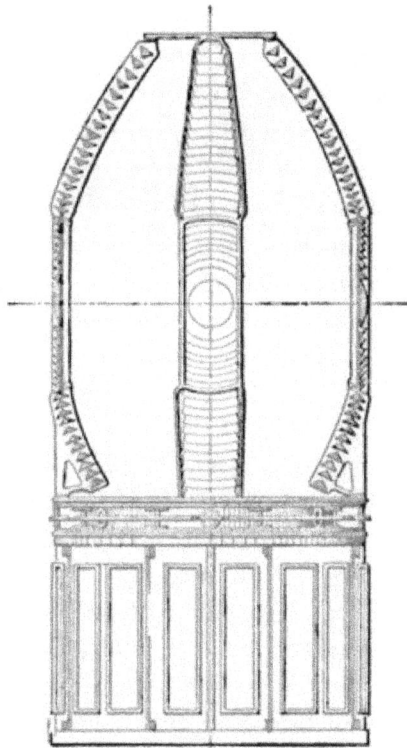

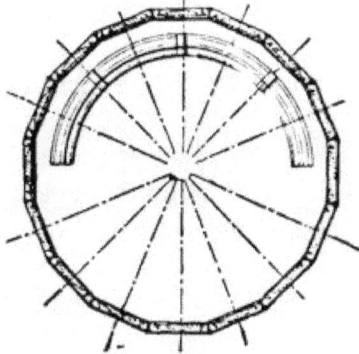

FIG. 78.

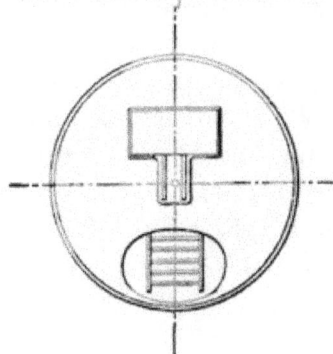

FIG. 80.

plane through the focus. It will be seen that a dioptric mirror is placed on the landward side of the arc. This mirror is arranged to form the image of the arc at one side of the carbons, so avoiding the interception of light which would result if the mirror were used in the ordinary way, and contributing to the horizontal divergence necessary. Further horizontal divergence is given by the form of the lens. In the ordinary revolving light the inner face of the lens is plane; here it is cylindrical, the axis of the cylinder being vertical. This method of obtaining horizontal divergence is a modification of a proposal of Mr. Thomas Stevenson,* M. Inst. C.E.; it is not mathematically accurate, inasmuch as the cylindrical form of the inner face of the lens not only displaces the emergent ray horizontally, but also, in the case of rays not in the vertical nor horizontal plane through the focus, to a small extent vertically; but the error is easily calculable, and is unimportant, provided the lens is narrow, and the horizontal divergence of the beam moderate. Fig. 79 shows a complete panel in elevation with revolving carriage. Fig. 80 shows the plan of the service table of the pedestal and lamp table. A new construction was adopted for the gunmetal framework of the optical apparatus to reduce the interception of light by the frame to a minimum. The metal segment *A*, Fig. 81, forms part of the lower prism frame, *B* part of the upper frame, while *C* and *D* are parts of the frame for the refracting portion of the apparatus; uprights *E* support the upper prism frames without throwing weight on the lens frames. With the ordinary con-

---

* " Lighthouse Construction and Illumination," p. 186.

structions of frame, Figs. 82 and 83, the equivalent of these ring segments A and B would intercept about double as much light as in this new construction.

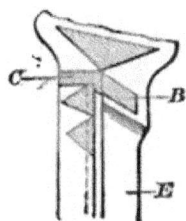
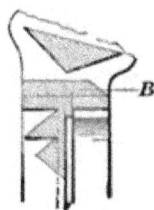
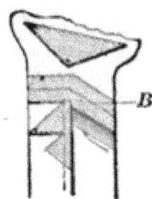
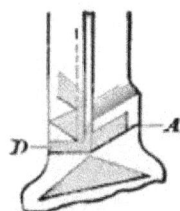
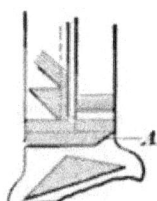
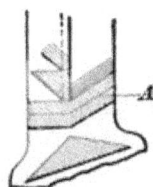

FIG. 81.   FIG 82.   FIG. 83.

Mechanism for Rotation.—The pedestal is similar to those designed by Sir James Douglass to permit the light keeper to obtain access from below to the interior of the apparatus without in any way interfering with its rotation. The clockwork is fitted with the governor, and maintaining power used by Messrs. Chance Brothers & Company for the last twelve years. The roller ring may be mentioned as of an improved type, for although it has been used for some years in all Messrs. Chance's lights, it has not been described before. The rollers and roller paths which carry the whole weight of the optical apparatus have long been made conical, so that the surfaces roll upon each other without twisting. There is consequently a very considerable radial force on each roller tending to force it outwards; the reaction against this force causes a very important part of the total frictional resistance. Fig. 84 shows a portion of the roller ring and one of the conical

 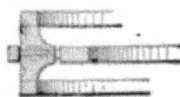 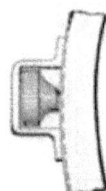

FIG. 84.   FIG. 85.   FIG. 86.

rollers, according to the old construction; Figs. 85, 86, according to the improved construction; in the former it will be observed that the thrust of the roller is received on a collar; in the latter, on the end of a pin. The reduction of friction is practically very considerable, and although of small importance in a slow-moving apparatus like Macquarie, is of great importance in heavier and quicker apparatus; for example, the triple flashing light at Bull Point, in Devonshire.

Lamps.—These are of the Serrin type, and were supplied by Baron De Meritens.

Lamp Table.—The arrangements for rapidly changing electric lamps, and for substituting gas or oil when desired, are shown in Figs. 87, 88, 89, and 90.

The intention was to use a gas lamp in clear weather, and half power or full power electric light in thick weather, according to the opacity of the atmosphere; but the author understands that in practice the electric arc is always used. The paraffin oil lamp is intended as a resource in case of failure of the supply of gas.

FIG. 87.

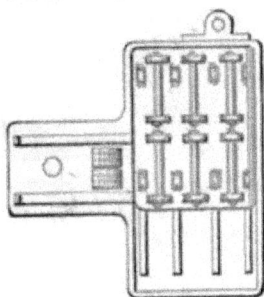

FIG. 88.

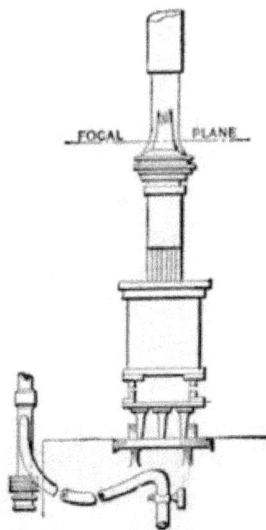

FIG. 89.
ARRANGEMENT FOR OIL LAMP.

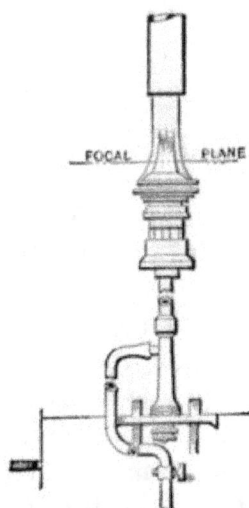

FIG. 90.
ARRANGEMENT FOR GAS BURNER.

Focussing the Arc.—Two approximately rectangular prisms are fixed upon the mirror frame at about 90° from each other, the longer face of each is plane, the other two faces convex, of such curvature as to form a good image of the arc upon the service table, as shown in Fig. 91.

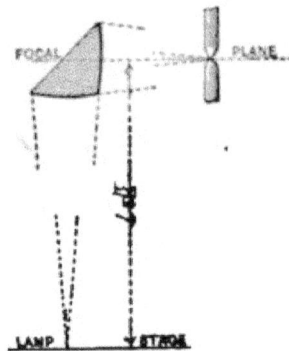

FIG. 91.

During daylight, a pointed sight or focimeter is placed at the position of the image formed by the lens of an object on the horizon; this then is the position which the arc should occupy. A sight is next taken over the focimeter into one of the adjusting prisms, and a bright object such as a threepenny piece placed on the service table, is moved about until its centre is seen in the prism, exactly upon the point of the focimeter; a mark is made in the then position of the object. When the arc is correctly adjusted, its image on the service table will be at the point where the mark is made. Two prisms are used in order to secure that the arc shall be in the centre of the apparatus as well as at the correct level.

Lantern.—The lantern is of the well known Douglass type.*

_____

* Minutes of *Proceedings of the Inst. C.E.*, vol. LVI., p. 77.

Dynamo-Electric Machines.—Two alternate current machines, with permanent magnets manufactured by De Meritens, were supplied. Each machine has five rings in its armature, and in each ring there are sixteen segments. In supplying one arc for a lighthouse the machine runs about 830 revolutions per minute, and gives a current of 55 ampères when half the coils are used, and of 110 when the whole of the machine is in action, the internal resistance in the two cases being 0.062 and 0.031 ohm. It is unnecessary to give a description of the machine as its general construction and dimensions are well known, but some numerical details are given below.

Engines.—Each machine is driven by an 8 h. p. Crossley gas engine through a belt without countershafting.

Tests.—Whilst the dynamo machines were at the works of Messrs. Chance Brothers & Company, a series of experiments was made in March, 1881, to determine their properties. The time is long passed when it would be profitable to give the details of these experiments, but the general conclusions drawn at the time are still interesting. When the external resistance was a metallic conductor with small self induction, it was found that with varying resistance and speed the currents observed agreed fairly

well with calculation from the formula $\dfrac{A}{\sqrt{R^2 + \left(\dfrac{2\pi\gamma}{T}\right)^2}}$*

in which R is the total resistance of the circuit, γ the self induction, and T the periodic time. When the machine

---

* Lectures on the "Practical Application of Electricity." Session 1882-83. Paper on "Some Points in Electric Lighting," reprinted in this volume.

was running 830 revolutions per minute $A = 67$ volts and

$$\left(\frac{2\,\pi\,\gamma}{T}\right)^2 = 0.197 \text{ in ohm squared, hence } \gamma = 6.4 \times 10^3$$

centimetres. The eighty sections of the machine are
arranged four in series, twenty parallel. For a single section the value of γ would be 32×10^8 centimetres. The
maximum induction in the core, which has an area of 5
square centimetres, is 24,600 or 4,920 per square centimetre. The loss of power was greater when the machine
was doing little or no external work than when that work
was great. This is clearly seen in the following table:—

Current ampères	7.70	73.60
Electrical work h. p.	0.69	5.66
Mechanical work applied	3.09	6.55
Loss	2.40	0.89

Photometric experiments were made upon the arc, and
simultaneous measurements of effective power applied and
of current passing. The red light was measured through
bright copper ruby glass, and the blue through a solution
of sulphate of copper and ammonia. The h. p. was measured by a transmission dynamometer; but the results must
be accepted with some reserve, on account of the difficulty
of ascertaining the mean tension in a strap which is constantly varying. The oscillations of the dynamometer
were damped by a dashpot containing tar.

	Half Power.	Full Power.
Red candles	1,968	4,708
Blue "	4,079	11,882
Current (ampères)	54 5	105
Mechanical power applied (h. p.)	4.5	6.9
Power expended in heating conducting wires (h. p.)	0 34	0.95

The results illustrate the fact that, as the current increases, the total light increases in a higher ratio, red light in a slightly higher ratio, and blue in a considerably higher.

The machinery for this lighthouse was sent out to New South Wales in November, 1881, and was put up and started under the superintendence of Mr. J. Barnett, the architect of the colony, to whom is mainly due the success of the whole from the first start. The glare of the light upon the sky is said to have been seen at a distance of over 60 miles, far beyond the distance at which it would cease to be directly visible. The only criticism from mariners has been that when somewhat near the lighthouse the flashes are so bright as to dazzle the eye. This is an excellent proof of the power of the light, as a much smaller proportion of the light is directed upon the nearer sea than in any previous lighthouse. The lesson is that with powerful electric lighthouses almost all the light should, in ordinary weather at least, be directed to the horizon, and that the quantity thrown upon the nearer sea must be strictly limited. This is only possible when the focal length of the apparatus is large.

TINO.

This station is on a small island at the mouth of the Gulf of Spezia. Fig. 92 is copied from the chart of the neighborhood. The focus is 386 feet above sea level. The distance of the sea horizon is 22.7 nautical miles, and the range practically 28 miles. The conditions, therefore,

were very similar to those of Macquarie, with the excep-
tion that it was required to throw some light down into
the channel between Palmaria and Tino. The lighthouse
itself presents some interesting historical features. The
buildings were originally a place of defence against the
pirates who occasionally made descents upon the coast.

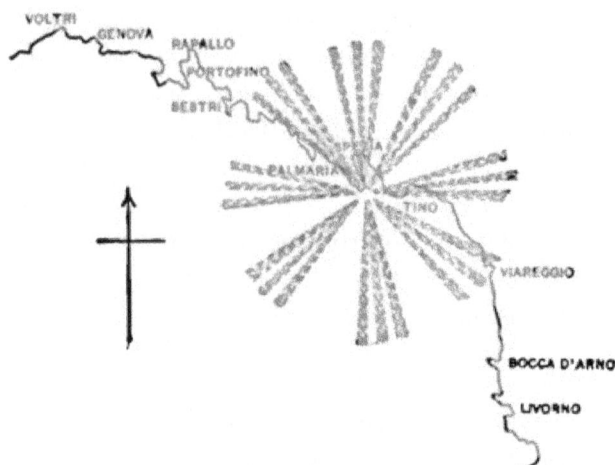

FIG. 92.—(Scale, 22 miles = 1 inch.)

Subsequently a coal fire lighthouse was established, and in
the spring of 1885 part of the stock of lignite was still
found to be in some of the buildings, where it had been
lying for fifty years. In 1839 a dioptric light was established, one of the earliest of Fresnel's types, the lens ring
being replaced by short straight prisms, which formed by
no means a bad approximation, and could be ground without special machinery. The present electric lighthouse has
been in contemplation for several years.

Optical Apparatus.—The distinctive character of the
light is a triple flash every half minute. The apparatus for
producing this effect is of the general form introduced by
the author in 1874. In October of that year he issued a
pamphlet pointing out the several advantages of group
flashing lights, showing for the first time a simple dioptric
apparatus suitable to their production, and also pointing
out how easy it is to give the group flashing effect with
catoptric apparatus. Since that time a large number of
dioptric group flashing lights have been made by Messrs.
Chance Brothers & Company, and some also in France,
and Mr. Allard has incorporated group flashing lights in
the system of distinctions he recommends; also a considerable proportion of the light vessels on the English coasts
have been converted into group flashing lights of the
catoptric system. On the ground of economy the second
order apparatus of 700 millimetre focus was adopted in the
case of Tino. It is just large enough for tolerably convenient service of the lamp by an attendant entering within the apparatus. The apparatus, shown in vertical
section in Fig. 93, and in horizontal section through the
focus in Fig. 94, has twenty-four sides, eight groups of
three; one group of three is shown in elevation in Fig. 95.
The horizontal divergence is obtained in exactly the same
way as at Macquarie, excepting that no mirror is used.
The metal framework, however, approximates to the
ordinary type, as the type used at Macquarie would have
been costly when applied to a triple flash light. The distribution of light vertically is as follows: upper and lower
prisms, and the central lens, with the two lens rings next
adjoining it, all to the horizon and most distant sea. The

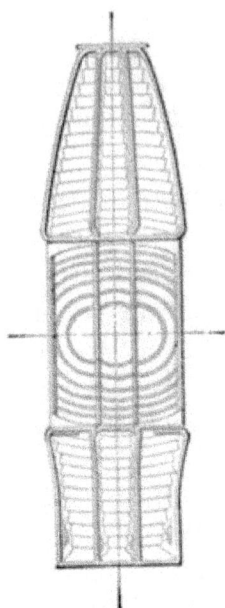

FIG. 95.

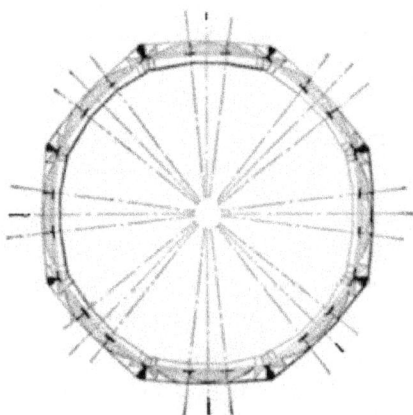

FIG. 94.

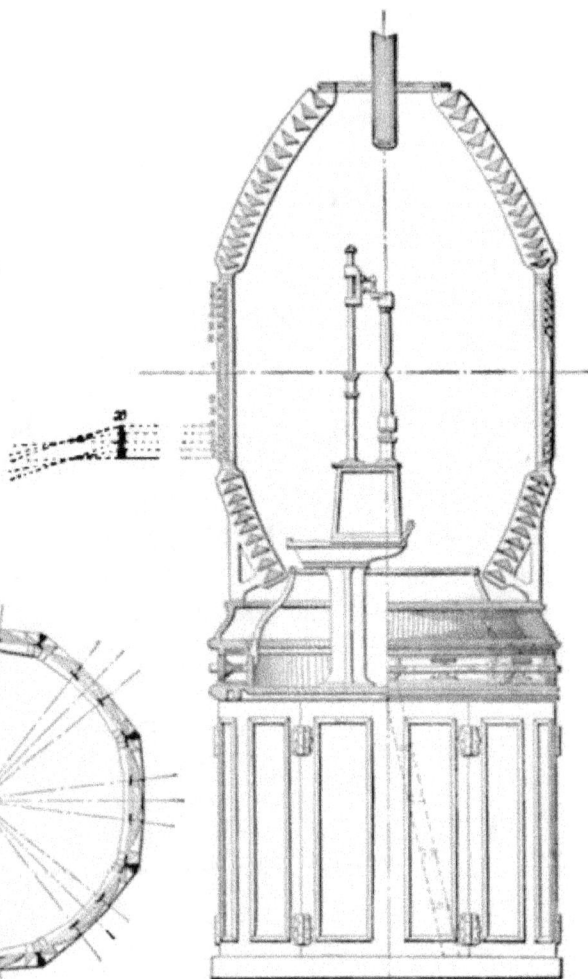

FIG. 93.

lens and lens rings direct their rays according to the following table, which is arranged in exactly the same way as the table already given for Macquarie:—

	I.	II.			III.		
		°	′	″	°	′	″
7 top...........		0	31	35	3	16	0
6 "		..			2	0	0
5 "		..			1	30	0
4 "		..			1	0	0
4 bottom...............................		..			0	45	0
5 "		..			0	30	0
6 "		..			0	30	0
7 "		all to the horizon.					

No. 7 bottom was directed wholly to the horizon, in order to avoid the horizontal bar of the lantern. It will be observed that the quantity of light thrown upon the nearer sea is much less in the case of Tino than in that of Macquarie, and that greater reliance is placed upon the accuracy with which the arc can be kept in focus; experience has justified these changes, as improvements of a perfectly safe nature.

A small part of each flash is bent downwards and distributed over the channel between Tino and Palmaria, by means of subsidiary prisms fixed upon the lantern, shown at *N*, Fig. 93. These subsidiary prisms are really superfluous, as the scattered light from the beams overhead is found to be as effective at this short distance. Fig. 96 shows the plan of lamp shunting table and service table.

Engines.—As there is no water upon the island, the practice of the Trinity House was followed, and two of the Brown hot air engines were supplied, each driving through a countershaft one of the machines. The countershafts could be connected by means of a Mather and Platt friction coupling, so that the two machines could be driven together, or either machine from either engine. Drawings of the Brown engines are given in Sir

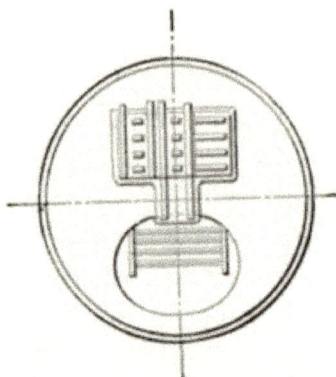

FIG. 96.

James Douglass's Paper.* The accompanying indicator diagrams Figs. 97 and 98.were taken from the compressing and working cylinders. Whilst these diagrams were taken, the effective power developed was measured by a friction brake on the driving pulley, and was found to be 9.1 h. p. Thus of 33.1 h. p. indicated in the working cylinder, 17.7 h. p. is employed in compressing the air, 6.3 h. p. is wasted in friction in various parts of the machine, and only 9.1 h./p. is effective upon the brake. The engines consume about 4 lbs. of coke per effective h. p. per hour. In future lighthouses, when a steam engine cannot be employed, it would be preferable on every ground to use gas engines, and manufacture on the spot either Dowson gas or

---

* Minutes of *Proceedings of the Inst. C.E.*, vol. LVII., Plate 6.

ordinary gas, according to the character of the fuel available.

Dynamo Machines.—There are two machines of exactly

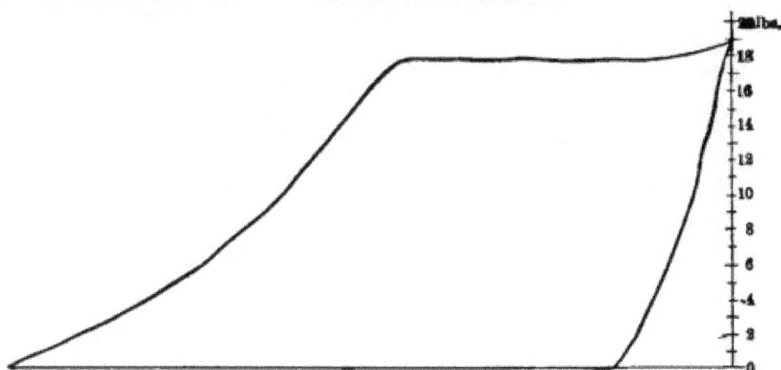

FIG. 97.—Compressor-pump cylinder, 24 inches in diameter.   Stroke, 22 inches. Indicated H. P., 17.7.

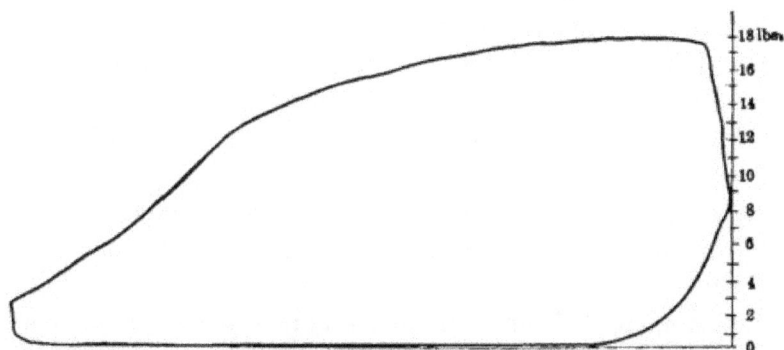

FIG. 98.—Working cylinder, 32 inches in diameter.   Stroke, 20 inches.   Indicated H. P., 33.1.   Revolutions per minute, 64.   Power on brake of fly-wheel, 9.12 H.P. Pressure in reservoir, 19 to 24 lbs.

the same type as those supplied for Macquarie, the only novelty lying in the method of using them.   In 1868 Mr. Wilde discovered, by experiment, that two alternate current dynamos, independently driven at the same speed, would, if electrically connected, so control each other's motions that they would add their currents. The author subsequently arrived at the same conclusion independently, on theoretical grounds, and gave a thorough explanation of the fact.* The result has been put to a practical application at Tino. The machines are connected to a single switchboard, so that each half of the two machines can at pleasure be connected to, or disconnected from, the main conductors. Thus a current can be supplied from either machine at half power, 55 ampères, or full power, 110 ampères, or from the two machines of double power, or about 200 ampères. Further, a change can be made without extinction of the light from one dynamo and engine to the other. Thus, suppose one machine is working full power, clutch the countershafts gradually together, so starting the second engine; throw on the band of the second machine, cut out half the first machine, and connect half the second machine at the switchboard; the two machines at once synchronize, without affecting the light. Disconnect the remaining half of the first machine, and connect the remaining half of the second, unclutch the countershafts, and stop the first engine. One man can effect the change, with no more disturbance of the light than a change from full to half power for about one second. A further conclusion, deduced from theoretical considerations, was that of two alternate current machines

---

* Lectures on the "Practical Applications of Electricity." Session 1882-83. Paper on "Some Points in Electric Lighting," reprinted in this volume. By Dr. John Hopkinson. And *Journal of the Society of Telegraph Engineers and Electricians*, vol. XIII., p. 496.

of equal potential, one could be used as a generator of
electricity, the other as a motor converting the current
generated back into mechanical power. It was found
impossible to verify this conclusion with such intermittent
driving as that of a hot air engine. But Professor W. G.
Adams effected the verification without difficulty at the
South Foreland, the motive power being steam.

Lamps.—These are the improved Serrin of Mr. Berjot.
One of the three lamps supplied is of larger size, for the
double power current from the two machines. This lamp
was said to be suitable for a still greater current, but with
about 200 ampères it soon became dangerously heated; a
simple modification rendered the lamp equal to the actual
work it had to do. It is, however, probable that for the
occasional circumstances when it is necessary to use so
great a current as 200 ampères in a lighthouse, a lamp
worked partly by hand would be preferable to a regulator
entirely automatic.

The apparatus was delivered in November, 1884, and
was put up by workmen from Messrs. Chance's workshops,
under the supervision of Mr. L. Luiggi, of Genoa, to whose
ability and energy the complete success of the lighthouse
is largely due. A complete test of the performance of the
light, as seen from the sea in all grades of its power, was
made in April, 1885, by a commission, consisting of Professor Garibaldi, of Genoa; Mr. Giaccone, engineer-in-chief
for Italian lighthouses; Captain Sartoris, and Mr. Luiggi,
the author attending on behalf of Messrs. Chance. The
light was well observed through rain, when distant 32
nautical miles, and although below the horizon, the position was precisely localized, and the triple flash distinction

unmistakable. At 18 miles distant the illumination of the flash upon white paper was sufficient to make out letters marked in pencil 1¼ inch high, and when 14 miles distant it was easy to ascertain the time from a watch. The light is frequently seen at a distance of 50 miles, near to Genoa.

A review of work which has been carried out naturally suggests many questions as to what conclusions experience has established, and what indications it gives of the probable direction for future developments. In the use of electric light in lighthouses, there are many questions upon which there is wide difference of opinion, questions both as to when and where electric light should be adopted, and questions as to the best way of employing it. It may not be unprofitable to allude to some of them. Although English engineers are now well agreed that a large optical apparatus should be used for the electric light, this opinion is not universally accepted. The advantages of a large apparatus have already been mentioned. To balance them, there is nothing on the other side but the less prime cost of the smaller apparatus. Although the difference of cost appears considerable when attention is confined to the optical apparatus, it is unimportant when the whole outlay on the lighthouse is brought into account. Cases are, however, conceivable in which a small optical apparatus such as a fourth order, having a focal distance of 250 millimetres, would be properly preferred; such, for example, as a harbor light which could be supplied with current from machinery also used for other purposes, but such cases are likely to be exceptional.

When a flame from oil or gas is the source of light, there
is of necessity a considerable divergence vertically; and the
distribution of the light through the angle of vertical
divergence is not at disposal, except to a very limited
extent in some cases, but is determined by the size and
character of the flame. With the electric arc and a large
optical apparatus it can be determined in considerable
measure how the light shall be distributed—how much
shall be sent to the distant sea, how much to the various
distances between the foot of the tower and a distance
of some miles. It becomes then a question what use is to
be made of this facility. The experience at Macquarie
and at Tino is emphatic, that it is in every way advantage-
ous to direct much the greater part of the light to the
horizon with a very small divergence, and to distribute the
comparatively small remainder over the nearer sea with intensity increasing with the distance.

A question allied to the last is this: Whether it be desirable to provide means of directing the strongest light
downwards on to the nearer sea in time of fog? The
answer must depend upon the circumstances of the particular locality. Take the case of a lighthouse on an
isolated rock, the purpose of which is primarily to be a
beacon to keep ships off that rock; a lighthouse which
would not exist were it not more practical or cheaper to
build and maintain the lighthouse rather than remove the
rock. Here surely it is of the greatest use to provide
means whereby, if the light cannot penetrate 2 miles, it
shall if possible be visible at 1 mile. But other cases
occur in which the lighthouse has to cover a long length
of coast, and has almost as much to do with points of the

coast 10 miles distant as with the point upon which it is placed, cases in which the lighthouse is far more useful in guiding the regular traffic passing within a radius of 20 miles or more than in preventing vessels running ashore within a mile of the tower. Such a light fails of its purpose if it can only be seen at a distance of a mile, covering less than $\frac{1}{100}$ part of its normal area of illumination; it becomes comparatively useless unless it penetrates. to something like its normal range, and its efficiency must be measured by the fewness of the occasions when it fails to do this. It is a grave question whether it be prudent in such cases to place upon the light keeper the responsibility of judging when the light should be dipped on to the nearer sea, the fact being that, if his judgment errs, he may actually diminish the range of the light, and cause unnecessarily the lighthouse to fail of fulfilling its most important function. It is easy for him to be misled if the fog is local and does not extend to any great distance from the lighthouse. Another element enters into the consideration—the height above the sea. If the focus be 100 feet above the sea level, the dip of the sea horizon is 9' 45'', and a ray dipping 9' 45'' below the sea horizon will meet the sea at a distance of 3.1 nautical miles from the tower. Even with a first order apparatus, if the arc be a powerful one, it is very difficult to render the light directed to the horizon from an elevation of 100 feet more powerful than that directed to a point distant 4 miles from the tower. Unavoidable divergence will render the two intensities practically equal.

Passing to questions of another class, what are the relative advantages in an electric lighthouse of continuous and alternating currents? Present practice tends altogether in favor of alternate currents, but this practice largely results from unfavorable experience of the older continuous current machines. These machines have in many respects been greatly improved in the last two or three years. The continuous current presents the advantage of greater economy of power in producing the current, less floor space required by the machine, and a smaller prime cost. The alternate current magneto machine, on the other hand, has the advantage that it may be driven with a defectively governed prime mover, with an indifferent lamp, and may suffer neglect with impunity; whereas the more compact and efficient continuous current machine would be in serious peril of destruction. Optical apparatus can be constructed suitable to make the most of either form of arc. Hot air engines have found favor for electric lighthouses, because in many cases there is no available supply of fresh water. The engines of which the author has experience are open to the objection that they take a great deal of room, are not economical of fuel, and do not govern so quickly as is desirable; the wear and tear also, when they are worked to anything like their full power, is very serious. A gas engine, with Dowson or other gas made on the spot, could be used with greater advantage.

Antecedent to all considerations as to the best apparatus and machinery to be used is the question under what circumstances, if at all, should electric light be used in a lighthouse? The Trinity House experiments at the South Foreland showed to demonstration that, where the issue to be decided was how to produce a light which should be capable of penetrating the furthest in all weathers, electric light could do that which could be done in no other way, and that it was the cheapest light of all when the price is estimated per unit of light. But the conclusion was also reached that an electric light must inevitably cost a large sum, both in first outlay and in maintenance; therefore that electric light is extravagant unless very extraordinary power is a necessity. This conclusion is doubtless a fair consequence of experience, but it is not an inherent property of electric light. Both the capital outlay and the cost of maintenance are greatly increased by the practice of so arranging the machinery as to provide, at all times, a light of very great power: whence it follows that the machinery must be placed at some distance from the lantern, and two men must always be on duty; one man in the lantern, and another with the machinery.

The essentials for a cheap electric lighthouse are, that for ordinary states of the atmosphere there shall be provided a plant under the easy control of the light keeper himself, which shall be precisely adapted to produce that amount of light which is wanted in ordinary states of the atmosphere; but for thick weather there shall be provided a much more powerful engine and dynamo, available also as a reserve in case the smaller machinery from any cause breaks down. The occasional machinery may be more remote from the lantern, as it is a small matter to require a second man to work on the comparatively rare occasions when the maximum power is needed. A small gas engine and a dynamo machine can be placed without any crowding in the room immediately below the lantern, and arrangements can be made whereby the light keeper, whether he is in the lantern or in the engine room, can ascertain at a glance whether the arc is in its proper position, with an error of less than 1 millimetre. The attendance on the lamp, rotating apparatus of the lens (if a revolving light), engine and dynamo, would be easy when the whole is brought together so as to be under observation at once; in fact the gas engine, dynamo, and lamp constitute together a gas burner which, though consisting of many parts, is automatic throughout, and requires nothing but the constant presence of a custodian, exactly as the gas lamp in a lighthouse requires a custodian as a guarantee against failure. The same end, viz., the concentration of the whole mechanical and electrical apparatus under one pair of eyes, could be attained, of course, in other ways. Accumulators could be used, or a petroleum engine.

In order to give definiteness and afford facilities for criticism, the better course will be to describe a suitable machinery; state what it will do, what attendance it will require, and what it will cost. The author proposes, then, for an electric lighthouse where small outlay is essential, the following: A Dowson gas producing apparatus and gas holder, the generator and superheater being in duplicate, each capable of making 1,200 feet of gas per hour, the gas holder having a capacity of 3,000 cubic feet.

An 8 h. p. nominal Otto gas engine and series wound dynamo machine, placed in a room near the base of the tower, and copper conductors to the lantern, the dynamo having magnet coils, divided into sections so as to supply a small current when required.

A 1 h. p. nominal Otto gas engine and dynamo machine, placed in the room immediately beneath the lantern floor, with gas pipe from the gas holder; three electric lamps, to receive either carbons 25 millimetres in diameter or any lesser size, with complete adjustments for accurate focussing; one paraffin lamp as a substitute; an optical apparatus of the second order of 70 centimetre focal distance. The cost of this apparatus would depend upon the character of the light it was intended to exhibit. To fix ideas, let it be assumed that the light is to be a half minute revolving light, showing all round the lighthouse. There could then be supplied a sixteen sided apparatus with pedestal and revolving machinery. Provision would be made in the optical apparatus for giving the horizontal and vertical divergence desired by the same methods successfully used in the lighthouses of Macquarie and of Tino.

Two focussing prisms would be fixed to form magnified images of the arc, on pieces of obscured glass let into the pedestal floor, so that the keeper, whether in the lantern or in the engine room, could see at a glance the state of the arc, and observe whether it is of proper length with the carbons in line, whether it is exactly at the right height and in the centre of the apparatus. An error of 1 millimetre would be glaringly apparent, and call for immediate adjustment, although its effect would be only a displacement of the beam 5' of angle.

The lantern would be 10 feet diameter, with bent plate glass.

The cost of the whole above described would be materially less than the cost of a first order light and lantern with oil lamp and large burners.

Now what result would be obtained? In fine weather the small engine would be used. Its effective power on the brake is fully $1\frac{1}{2}$ h. p.; from this $1\frac{1}{4}$ h. p. the dynamo

machine produces considerably over 800 watts, say 800
watts in the arc itself, or 20 ampères through a fairly long
arc of 40 volts. Of course the value of this in candles depends upon the color in which it is measured, and the
direction in relation to the axis of the carbons. In red
light the mean over the sphere would certainly exceed
1,200 candles. In clear weather or in slight haze or rain,
the beam of this light through the lenses would be much
more powerful at the horizon and on the more distant sea
than any single focus light with oil or gas as the illumi-
nant, and would at least be fairly comparable with anything yet exhibited with oil or gas whether triform or
quadriform.   But on the nearer sea the illumination
would be reduced, so that no annoyance would be caused
by dazzling flashes. In thick weather or indeed in any
weather when there was a doubt as to the visibility at the
horizon of the lower power, the large engine would be
used under the superintendence of the second keeper.
This engine will give 10 h. p. on the brake, and there is no
difficulty in obtaining 85 per cent. of this as useful
electrical energy outside the machine, that is, 6,340 watts.
From this deduct 10 per cent. for the leads and the lamp
and for steadying the arc, leaving 5,710 watts in the arc
itself, or 114 ampères, with a difference of potential of 70
volts. Having regard to the fact that the optical apparatus here proposed acts upon a larger portion of the
sphere than that used in the South Foreland experiments,
that the vertical divergence is less, and that the potential
difference is greater and the current continuous, although
less in quantity, it may safely be assumed that the power of
the resulting beam would not be inferior. It hence follows,

from the South Foreland experiments, that in any fog the flashes would penetrate farther than those of any existing gas or oil light. The increased size of crater, compared with that produced by the current of 20 ampères, will give increased vertical divergence, and so cause the maximum illumination to be attained at a less distance from the lighthouse. The attendance of two men would suffice for all the duties of the lighthouse, because under ordinary circumstances one man only need be on duty excepting for two to three hours while gas is being made. The consumption of coal would be 4 lbs. per hour of lighting, of water about ½ gallon, of carbons about 4 inches. The whole cost of maintaining the light would differ little from that of an ordinary oil light of the first order.

Though it be the fact, that it is possible to exhibit an electric light at moderate cost, it does not follow that it is suitable for all ocean lights. There is no room in a rock lighthouse tower for a gas plant, and few would at present be prepared to recommend a petroleum engine burning oil of a low flashing point. The light keeper again must understand a gas producer, a gas engine, a dynamo, and an arc lamp, instead of only a paraffin lamp and burner, and arrangements must exist for repairing the more extensive machinery. Such considerations will justly weigh against the use of the electric light in remote stations and in countries where the labor available is not capable of much training.

It may possibly be said that in this paper no definite conclusions are reached as to whether electricity or some other agent is the best source of light in lighthouses

generally, nor yet, if electricity be adopted, what is the
best way of producing the light and optically dealing with
it.   The answer is that it is impossible on many points to
arrive at general conclusions.   Each case must be judged
according to its special circumstances.

IF YOU WISH TO KNOW

The latest and best works on the principles and theory of Electricity, or relating to any particular application of Electricity, THE ELECTRICAL WORLD will be pleased to promptly furnish the information, personally or by letter, free of charge. If you live in or near New York, and would like to examine any Electrical Books, you are cordially invited to visit the office of THE ELECTRICAL WORLD and look them over at your leisure.

Making a specialty of Electrical Books, there is no work relating directly or indirectly to Electricity that is not either published or for sale by The Johnston Company, and the manager of the Book Department keeps himself at all times familiar with the contents of every work issued on this subject at home and abroad.

Any Electrical Book in this catalogue, or any Electrical Book published, American or foreign, will be mailed to ANY ADDRESS in the world, POSTAGE PREPAID, on receipt of price. Address and make drafts, P. O. orders, etc., payable to

THE W. J. JOHNSTON COMPANY, Ltd.,

TIMES BUILDING, NEW YORK.

EXPERIMENTS WITH

ALTERNATE CURRENTS

Of High Potential and High Frequency.

By NIKOLA TESLA.

Cloth.   156 pages, with Portrait and 35 Illustrations.   Price, $1.00.

This book gives in full Mr. Tesla's important lecture before the London Institution of Electrical Engineers, which embodies the results of years of patient study and investigation of the phenomena of Alternating Currents of Enormously High Frequency and Electromotive Force.

The book is well illustrated with 35 cuts of Mr. Tesla's experimental apparatus, and contains in addition a biographical sketch, accompanied by a full-page portrait, which forms a fitting frontispiece to a lecture which created such widespread interest.

Every Electrician, Electrical Engineer or Student of Electrical Phenomena who makes any pretensions to thorough acquaintance with recent progress in this important field of research which Mr. Tesla has so ably developed, must read and reread this lecture.

Copies of this or any other electrical book or books published will be promptly mailed to any address in the world, POSTAGE PREPAID, on receipt of price. Address

THE W. J. JOHNSTON COMPANY, Ltd.,

TIMES BUILDING, NEW YORK.

AN IMPORTANT NEW BOOK.

ALTERNATING CURRENTS,

Treated Analytically and Treated Graphically.

BY

FREDERIC BEDELL, Ph.D., and A. C. CREHORE, Ph.D.,

(Cornell University.)

Uniform in size and style with " The Electric Railway in Theory and Practice," by O. T. Crosby and Dr. Louis Bell.

Cloth.   300 Pages and 112 Illustrations.   Price, $2.50.

While there are many monographs and special treatises on alternating currents, they are either fragmentary or special in character, or couched in mathematical language requiring a special mathematical education to interpret.

In this volume the theory of alternating currents is, for the first time, treated in a connected and logical manner, and in mathematical language familiar to the ordinary mathematical public, while the graphical extension can be followed by those not having a special knowledge of mathematics.

Some parts of this volume have been published in separate papers, and from the cordial welcome they received it is believed that the present work will fill a distinct want in an important branch of electrical science.

Mailed prepaid to any address on receipt of the price by the Publishers,

THE W. J. JOHNSTON COMPANY, Ltd.,

TIMES BUILDING, NEW YORK.

Publications of the W. J. Johnston Co., Ltd.

THE ELECTRICAL WORLD. An Illustrated Weekly Review of Current Progress in Electricity and Its Practical Applications. Subscription, in advance, one year.... **$3.00**

JOHNSTON'S ELECTRICAL AND STREET RAILWAY DIRECTORY, Containing Lists of Central Electric Light Stations, Isolated Plants, Electric Mining Plants, Street Railway Companies—Electric, Horse and Cable—with detailed information regarding each; also Lists of Electrical and Street Railway Manufacturers and Supply Dealers, Electricians, etc. Published annually.. **5.00**

THE TELEGRAPH IN AMERICA. By Jas. D. Reid. 894 royal octavo pages, handsomely illustrated, Russia... **7.00**

DICTIONARY OF ELECTRICAL WORDS, TERMS AND PHRASES. By Edwin J. Houston, A.M. Second Edition, entirely rewritten. 5,000 definitions, 562 double column octavo pages, 570 illustrations........... **5.00**

THE ELECTRIC MOTOR AND ITS APPLICATIONS. By T. C. Martin and Jos. Wetzler. With an appendix on the Development of the Electric Motor since 1888. By Dr. Louis Bell. 315 pages, 353 illustrations........ **3.00**

THE ELECTRIC RAILWAY IN THEORY AND PRACTICE. The First Systematic Treatise on the Electric Railway. By O. T. Crosby and Dr. Louis Bell. Octavo, 400 pages, 179 illustrations.... .... **2.50**

ALTERNATING CURRENTS. Treated Analytically and also Treated Graphically. By Frederick Bedell, Ph.D., and Albert C. Crehore, Ph D. Uniform in size and style with Crosby & Bell's "Electric Railway".. **2.50**

PRINCIPLES OF DYNAMO-ELECTRIC MACHINES and Practical Directions for Designing and Constructing Dynamos. By Carl Hering. Sixth thousand. 279 pages, 59 illustrations. **2.50**

ELECTRIC LIGHTING SPECIFICATIONS for the Use of Engineers and Architects. By E. A. Merrill. 175 pages.. **1.50**

THE QUADRUPLEX. By Wm. Maver, Jr., and Minor M. Davis. With Chapters on Dynamo-Electric Machines in Relation to the Quadruplex, Telegraph Repeaters, the Wheatstone Automatic Telegraph, etc. 126 pages, 63 illustrations.............................. **1.50**

THE ELEMENTS OF STATIC ELECTRICITY, with Full Description of the Holtz and Töpler machines. By Philip Atkinson, Ph.D. 228 pages, 64 illustrations.... **1.50**

LIGHTNING FLASHES. A Volume of Short, Bright and Crisp Electrical Stories and Sketches. 160 pages, copiously illustrated........................... **1.50**

ELECTRICITY AND MAGNETISM. A Series of Advanced Primers. By EDWIN J. HOUSTON, A.M. 297 pages, 116 illustrations....................... $1.00

RECENT PROGRESS IN ELECTRIC RAILWAYS. Being a Summary of Current Advance in Electric Railway Construction, Operation, Systems, Machinery, Appliances, etc. Compiled by CARL HERING. 386 pages, 120 illustrations.................................... 1.00

ORIGINAL PAPERS ON DYNAMO MACHINERY AND ALLIED SUBJECTS. Authorized American Edition. By JOHN HOPKINSON, F.R.S. 249 pages, 98 illustrations................................ 1.00

DAVIS' STANDARD TABLES FOR ELECTRIC WIREMEN. With Instructions for Wiremen and Linemen, Rules for Safe Wiring, Diagrams of Circuits and Useful Data. Third Edition. Revised by W. D. WEAVER. 1 00

UNIVERSAL WIRING COMPUTER for Determining the Sizes of Wires for Incandescent Electric Lamp Leads and for Distribution in General Without Calculation, with Some Notes on Wiring and a Set of Auxiliary Tables. By CARL HERING. 44 pages.................. 1.00

EXPERIMENTS WITH ALTERNATING CURRENTS OF HIGH POTENTIAL AND HIGH FREQUENCY. By NIKOLA TESLA. 146 pages, 30 illustrations... 1.00

LECTURES ON THE ELECTROMAGNET. Authorized American Edition. By Prof. SILVANUS P. THOMPSON. 287 pages, 75 illustrations 1.00

PRACTICAL INFORMATION FOR TELEPHONISTS. By T. D. LOCKWOOD. 192 pages...................... 1.00

WHEELER'S CHART OF WIRE GAUGES.......... 1.00

PROCEEDINGS OF THE NATIONAL CONFERENCE OF ELECTRICIANS IN PHILADELPHIA. 300 pages 23 illustrations............................ .75

WIRED LOVE: A Romance of Dots and Dashes. By ELLA CHEEVER THAYER. 256 pages.................. .75

HERING'S TABLES OF EQUIVALENTS OF UNITS OF MEASUREMENT............................. .50

Mailed prepaid to any address on receipt of the price by the Publishers

THE W. J. JOHNSTON COMPANY, Ltd.,

TIMES BUILDING, NEW YORK.

The Pioneer Electrical Journal of America.

Read wherever the English Language Is Spoken.

THE ELECTRICAL WORLD

IS THE

Largest, Most Handsomely Illustrated, and Most Widely Circulated Electrical Journal in the World.

It should be read not only by every ambitious electrician anxious to rise in his profession, but by every intelligent American.

The paper is ably edited and noted for explaining electrical principles and describing new inventions and discoveries in simple and easy language, devoid of technicalities. It also gives promptly the most complete news from all parts of the world relating to the different applications of electricity.

Subscription, including Postage in the U. S., Canada or Mexico, $3.00 a Year.

May be ordered of any Newsdealer at 10 cents a week.

THE W. J. JOHNSTON COMPANY, Ltd.,

TIMES BUILDING, NEW YORK.

www.ingramcontent.com/pod-product-compliance
Lightning Source LLC
Chambersburg PA
CBHW021523210326
41599CB00012B/1367